MARKETING YOURSELF WITH TECHNICAL WRITING

A GUIDE FOR TODAY'S PROFESSIONALS

WILLIAM M. VATAVUK

LEWIS PUBLISHERS
Boca Raton   Ann Arbor   London   Tokyo

Library of Congress Cataloging-in-Publication Data

Vatavuk, William M., 1947-
  Marketing yourself with technical writing : a guide for todays professionals / William M. Vatavuk.
      p. cm.
  Includes bibliographical references and index.
  ISBN 0-87371-478-4
  1. Technical writing—Handbooks, manuals, etc.   I. Title.
T11.V39   1992
808'.066—dc20   92-17217
                                                                      CIP

LEWIS PUBLISHERS
121 South Main Street, Chelsea, MI 48118

PRINTED IN THE UNITED STATES OF AMERICA
1 2 3 4 5 6 7 8 9 0

Preface

"Why should I write technical papers (or articles or books)? What's in it for me?" If you've ever wondered why, you're not alone. Many engineers, scientists, physicians, accountants, attorneys, and other professionals — privately, publicly, and self-employed — have asked themselves similar questions. Perhaps only academics, whose advancement and job security are inextricably tied to how often and where they get published, can answer them definitively.

But what if you're not in academia? How can getting published help you — professionally, financially, and psychically? This book tells you how. It also answers such questions as: "Who will publish this article I've just written?" and "I have a great book idea. How do I go about writing it and getting it published?" To help guide you through the complex (and sometimes frustrating) world of publishing, this book includes advice on selecting your writing topic, contacting and querying prospective periodicals and publishers, organizing your article or book, and getting it into print.

Also in these pages is rarely compiled information on the technical publications market — periodicals and publishers alike — as well as valuable tips on the legal and business aspects of publishing. The latter include copyrights, contracts, and taxes. And for inspiration, you'll read what some very successful technical writers, editors, and publishers have to say about writing and publishing in today's competitive marketplace.

Indeed, the theme of this book is *marketing*, as my primary aim is to help you market your technical writing, to make it more publishable. For no matter how brilliant and well written your work is, it's of limited value unless you get it into print.

William M. Vatavuk
Durham, North Carolina
March 14, 1990 to May 16, 1992

The Author

William M. Vatavuk is a senior chemical engineer with the Office of Air Quality Planning and Standards (OAQPS), U.S. Environmental Protection Agency, where he has been employed since 1970. For 17 of those 22 years, he has specialized in air pollution control cost analysis, work that has supported the setting of national emission and air quality standards. A recognized international expert in this field, he has published one book (*Estimating Costs of Air Pollution Control*, Lewis/CRC Press, 1990) and over 40 technical articles on cost analysis-related topics for such magazines as *Chemical Engineering* and the *Journal of the Air and Waste Management Association*. He has also received two EPA bronze medals for his work in this area.

Also, since 1975 he has been a part-time freelance writer. During that time, he has published hundreds of articles and reviews for many trade periodicals, ranging from *Catholic Digest* to *Writer's Digest*. He has also written *Dawn of Peace*, a popular history book that deals with a little-known, but important Civil War event. This book was nominated for the 1990 Pulitzer Prize for American History.

A graduate of Youngstown State University (Bachelor of Engineering, 1969), he is a registered professional engineer in North Carolina and a charter member of the OAQPS Human Resources Council. Listed in the *Who's Who Environmental Registry*, he has also been nominated for *Marquis – Who's Who in the South and Southwest*. Mr. Vatavuk lives with his wife and son in Durham, North Carolina.

Acknowledgments

First, I sincerely want to thank six very special people—the writers, editors, and publishers who contributed their advice and opinions to Chapter 7: Robert W. Bly, Harold Englund, Jeffrey Hillier, Roy Meador, Henry Petroski, and Dick Young. Without their inspirational input, this important chapter couldn't have been written. That much is obvious. What's not so evident is that their contributions gave this book an extra dimension, adding color and concreteness to what is, in essence, an abstract subject. Teaching someone how to get his/her writing published is one thing. Getting others to show and tell how they have done so successfully means so much more.

I'm also "much obliged" to the editors and publishers who contributed invaluable data on today's technical writing market—information that forms the nucleus of Chapter 5. Their names and affiliations are listed in Table 5.2. Of them, I'm especially grateful to Mark Rosenzweig and Diane Pirocanac, who really "went the extra mile" in helping me compile this information.

Conventional wisdom notwithstanding, no writer writes in a vacuum. He or she can and will benefit from the occasional pat on the back or (if appropriate) kick in the posterior. Brian Lewis, my publisher, Kathy Walters, my acquisitions editor, and Keith Doyle, my project editor, fulfilled this supportive role very well. They offered many helpful suggestions and kind words of encouragement while this book gradually took shape in my hands. To them, *merci beaucoup!*

Last, but by no means least, I must thank my wife, Betsy, for her love and unswerving support during the entire book-making process, from query to page proofs. Without her indispensable help, you wouldn't be reading this now.

Dedication

This book is dedicated to my biggest fan, my toughest critic, and my best buddy — who are one and the same person: my son, William Chandler Vatavuk.

Disclaimer

This book was written by William M. Vatavuk in his private capacity. No official support or endorsement by the Environmental Protection Agency or any other agency of the federal government is intended or should be inferred.

Contents

1   Why *You* Should Write for Publication

*Reading maketh a full man, conference a ready man, and writing an
exact man.*
Francis Bacon, *Of Studies*

Nearly every subject known to humanity has been written about at
one time or another. For it is our nature to record our findings, thoughts,
feelings — to preserve them for others, both for now and the future.
Nevertheless, there is, and always will be, just two kinds of writing:
published and unpublished. Though not necessarily superior to the
unpublished, published writing is invariably read by more and, consequently, has a much more profound impact on our world. This is as
true of technical writing as it is of any other kind, fiction or nonfiction.
Unfortunately, many technical professionals shy away from getting
their work published, reasoning that: (1) they can't write, (2) writing
takes too much time and effort, (3) there's no "payoff" to writing,
financial or otherwise, or (4) all of the above. The primary purpose of
this book is to remove these mental and emotional roadblocks from
potential writers' minds, and to show that getting your writing published
is worth the effort.

WHAT IS "TECHNICAL WRITING"?

First, let's define "technical writing". A working definition might be:
"Technical writing is writing that deals with strictly delineated, rigor-
ously founded subject matter specific to a particular field or discipline.
Moreover, it is generally written for, read by, and published in litera-

ture supported by practitioners of that discipline." Let's disassemble that definition. "Strictly delineated" means that the subject matter is bounded by the length, width, and depth of the discipline (or subdiscipline). For instance, chemical engineering (Ch.E.) is an established field that most professionals (and much of the general public) recognize by name. However, many subspecialities crowd under the Ch.E. umbrella: heat and mass transfer, thermodynamics, reaction kinetics, and process dynamics, to name just four. Each of these subspecialities has well-defined boundaries that are rarely broached. A reaction kinetics expert would have little interest in, or use for, a revolutionary procedure for designing distillation columns (though he *should* have at least a working knowledge of column design). "Rigorously founded" means that the subject matter is structured around some set of physical, psychological, or social laws or principles. Economics, for example, is not a science in the strict sense of the word, as it is not based on any physical laws. Nonetheless, economics is founded upon the *predictability* of human behavior in the marketplace. Illustration: If item A costs more than item B, and a rational purchaser deems both to be of equal value, he/she will buy item B. Put technically, a purchaser always tries to "maximize his/her marginal utility". Economists have codified these predictable preferences into laws, such as the Law of Supply and Demand, the cornerstone of the field.

In any event, writing that meets the above criteria can qualify as "technical". Moreover, this writing will usually appear in a periodical or book written for and read by professionals who specialize in the discipline in question.

TECHNICAL WRITING AND PROFESSIONAL DEVELOPMENT

As this definition implies, technical writing is hard work. That very fact is perhaps the biggest stumbling block to technical professionals who would otherwise write that article, paper, or even book. But what they fail to realize is that technical writing and publication can provide a hefty boost to one's career.

There are several ways to demonstrate professional development. Earning an advanced degree is one. So is participation in professional organizations. Both provide sound, well-recognized proof of career growth. And the latter is an excellent way to develop contacts that can prove to be invaluable later on. Yet another is presenting technical papers at regional and national conferences. This is a good way both to make contacts and demonstrate technical competence.

A fourth and, in my opinion, the most tangible measure of professional growth is technical publication. It offers several advantages that the others either do not or cannot provide. The first one is *widespread exposure to your professional peers.* If you earn that M.S. or M.B.A., who else is going to know it, aside from your family, colleagues, friends, and recipients of your resume? Similarly, if you present a paper at a convention of your colleagues, how many will hear it? A few hundred at most? Even if your paper were to be reprinted in the conference symposium, how many more would read it? After all, these symposia usually enjoy at most a limited circulation.

Suppose that instead of presenting that paper you had submitted it to a large-circulation journal instead? (Of course, you can do both: there's nothing to prevent you from presenting the paper *and* then getting it published.) The potential readership would be orders of magnitude higher. For instance, the readership of *Chemical Engineering* is nearly 300,000 worldwide. Thus, your name, professional credentials, and (in many cases) your photograph would be presented to thousands of your colleagues and many others outside your field. In one such article, you can reach more of your peers than through a score of conferences.

Secondly, your article allows you to showcase not only your expertise, but *yourself,* as well. For your writing style — your sentences and paragraphs, your vocabulary, your tone — reflect *you* in as many ways as your author bio and photo. What better way to sell yourself?

Finally, publication affords a *permanence,* an immortality if you will, that neither a graduate degree nor a professional association can offer. Decades from now will they remember that excellent paper you presented? But that journal article you wrote, or better yet that book, will be remembered. Most large technical libraries and the Library of Congress will still have a copy of it. And who knows? Some graduate student or researcher may happen upon it, read it, and glean something useful from it. In this way, publication allows us to influence others long after we've gone. In a sense, it sends forth an invisible cord that stretches "from here to eternity".

THE BUSINESS OF WRITING

Technical writing provides more than a means to professional growth or to establishing a literary legacy. It offers something more tangible — *money.* One might argue that money does not (nor cannot) motivate professionals to write for technical periodicals. After all, these publications usually pay little or nothing for contributions. In some cases, their circulations are so low that they cannot generate advertising revenues large enough to allow them to pay much to contributors. Other publications adopt the somewhat patronizing attitude that, because professionals already earn healthy incomes from their employers or businesses, the money they would receive for an article would be comparatively trivial. Still others, though they enjoy hefty advertising income, refuse to pay contributors, simply because they are tight-fisted. Both the patronizing and the stingy publications hide their true positions behind such statements as: "Professionals write for recognition, not for money."

Fortunately, not all technical periodicals and publishers share this attitude. As Chapter 5 shows, a number of periodicals *do* pay contributors reasonable "honoraria". These payments range from approximately $10 per column-inch for technical news pieces to $350 for feature articles. Though these fees are by no means exorbitant, they do help defray authors' costs (typing, photocopying, postage, etc.) and provide some return for their time and effort.

Even a well-paying periodical pays only once for a given contribution. A book, however, keeps paying an author as long as it keeps selling. How much? That depends on the book, the publisher, the author, and market forces outside their control. Let's take a close look at each of these elements.

The Book

First, for a book to sell a healthy number of copies, it must, first of all, contain technical information that a large number of readers would find not only useful, but essential to have. Ideally, this information should make their work easier to perform, while increasing their knowledge. This doesn't mean that readers have to have an immediate application for the information. They may also want the book for use as a reference. Handbooks like the *CRC Handbook of Chemistry and Physics* and the *Physician's Desk Reference* are examples of popular (and healthy-selling) books.

Next, the information needs to be organized and presented in such a way that the reader not only finds the book easy to use but a pleasure to pick up, time and again. No matter how useful the technical information is, if a reader can't find it in the book or interpret it once he finds it, the book is of little use. Moreover, the physical attributes of the book — cover, typeface, layout of tables and figures, and feel of the paper — must also be appealing. Not long ago, I happened upon an air pollution control book published in the Soviet Union. Since it was written in Russian, I understood very little of it, except for a few engineering equations. But what struck me most was the *appearance* of the book. The cover was a shoddy cloth material mounted on a flimsy cardboard backing, the binding was glued (not sewn), the paper had the consistency and strength of week-old newsprint, the illustrations were crudely drawn, and the type was microscopic. Even if this book were written in English, I would have found it very difficult and unpleasant to use, no matter how accurate and timely the information therein.

Thirdly, the book should carry a reasonable price tag. Most readers would balk at buying an over-priced book, even if it contained information essential to them. They would, instead, wait until the book was available in the firm's library or borrow a copy from a more affluent colleague.

The Publisher

Without a good publisher to support it, even the best book will wither on the vine. Why is a publisher important? As we'll see in Chapter 4, a technical publisher performs many functions. First, he accepts the draft book ("manuscript") from the author, sends it out for "peer review", compiles the review comments, and sends them to the author. Once he receives the revised script from the author, the publisher assigns it to an editor who then consults with the author as he guides the book along the road to publication: editing, preparation of galleys/page proofs, selecting typeface, designing the cover, writing the index, and other tasks.

A competent publisher does all of these things well; however, the difference between a competent publisher and a superior publisher can be summed up in one word: *marketing*. There's a saying among professional golfers: Just about every golfer on the P.G.A. tour can compete with the leading money winners from tee to green. But, what separates the journeymen from the masters is how well they play once they pull out their putters. *Marketing is to book publishing as putting is to golf.*

Marketing itself covers several different, but interrelated tasks: designing and distributing advertising flyers and book catalogs, mailing copies of the book to reviewers, paying sales calls to bookstores, making presentations at trade fairs and conventions, and, in general, doing as much as possible to ensure that as many people as possible learn about the publisher's books. Some publishers concentrate on direct-mail advertising, while others focus their sales on bookstores. Yet others prefer to sell through book clubs and other media. (However, most publishers use all of these outlets, to a greater or lesser extent.) Verily, aggressive marketing can significantly boost a book's sales.

So can a long "back list". A back list is a publisher's term for the list of titles that are still in print months, even years after they were first published. Back lists are typically very short in the "trade" publishing industry — firms who publish popular fiction and nonfiction. With the exception of such giants as Stephen King, the books of most popular authors are kept in print for, at most, 1 or 2 years, after which they are "remaindered" — sold at huge discounts to wholesalers who often resell them to discount stores, drugstores, and other retailers.

Technical publishers take a much longer profit view. As a rule, they keep their titles in print for several years, thus ensuring continued sales revenue for them and royalties for their authors. For example, the publisher of this book (Lewis Publishers, a CRC Press subsidiary) has kept nearly all of its books in print since it first started publishing in 1984.

The Author

Other than writing the best book he can and delivering it to his publisher on time, what can an author do to help his book sell? First, he can cooperate with his editor in making requested manuscript changes, reviewing galleys and/or page proofs, and preparing the index. An author should be open to the editor's suggestions, especially those that would make the book easier to read and use. After all, the editor is a technical professional whose judgment in such matters should be sounder than the author's. The author can also help to market the book by suggesting periodicals in which it can be reviewed, by writing journal articles based on it, and by taking advantage of other opportunities to publicize it. (The author-editor-publisher "partnership" will be discussed further in Chapters 3 and 4.)

The Marketplace

As we've just seen, the author and publisher can do a lot to boost book sales. But try as they might, there is not much they can do about the marketplace. It is totally out of their control. The impact of market forces on a book's sales can range from imperceptible to impressive. If a book is released at the right time, it may sell like proverbial hotcakes. If it's not, it may just as easily gather dust in the publisher's warehouse.

It's also true that the impact of market forces on the sales of a given book (or even class of books) is very difficult to measure. A publisher can, perhaps, survey the buyers to ascertain what prompted them to buy this or that book, what they plan to use it for, and so forth. But such surveys are expensive to conduct and, unless the sample size is sufficiently large, of dubious validity.

So, employing equal parts science and artistry, publishers endeavor to "read" the collective book-buying mind to predict just what sorts of books will be big sellers during the next year or two. (Given the lead time needed to write and publish a book, a year is the shortest time horizon to gaze at.) This reading is based mainly on sales experience. By examining past sales records, a publisher can see if there have been any patterns in book buying. For example, a book on preparing personal income taxes would likely sell very well between January 1 and April 15, but not so well during the rest of the year. Similarly, a book on the design of solar collectors would sell the most copies soon after Congress passed legislation giving tax credits to homeowners that install these systems.

OF DRUDGERY AND DEADLINES

However, there is a reverse side to the writing coin. First, as we said earlier, technical writing is *hard work*. Inasmuch as it has to present complex ideas and concepts clearly and comprehensively, technical writing requires more attention to detail than, say, the writing of a romance novel. To this, you might reply: "So what else is new? Every time I write a monthly progress report, I do the same thing. Draft and redraft, then make more changes once my supervisor sees it. You'd think I was writing the Gettysburg Address, not some report that hardly anyone ever reads!"

You're absolutely right; but if you think that composing a monthly report is hard, try writing a technical article or book. Since an article or a book will be read by so many more, and because your professional reputation will be on the line, you need to write with more care and *pain*.

However, you needn't feel alone. Many of the best known, most successful writers find writing to be onerous work. In the late 1940s, Robert Van Gelder wrote a book entitled *Writers and Writing,* for which he interviewed writers about their craft.[1] Some of their responses are veritable gems that have withstood the test of time. "Most of it is tough going," said Ernest Hemingway. "The work never gets any easier," echoed poet Robert Nathan, author of "Portrait of Jennie" and other works. Sinclair Lewis said it best: "Writing is just work — there's no secret. If you dictate or use a pen or type or write with your toes — it is still just work."

Over 30 years later, horror writer Dean R. Koontz (author of *Hideaway, Midnight,* and other bestsellers) wrote in his *How to Write Bestselling Fiction:* "Don't get the idea that writing is easy work. It is the hardest thing I have ever done. In my younger days, I held a few jobs that involved heavy physical labor…(but those) grueling job(s) never tired me out as completely as writing does."[2] As we'll see in Chapter 7, some technical writers also find writing to be demanding. Yet they also find it to be most rewarding work, psychically as well as financially.

Why is writing so difficult? Why is it the mental equivalent of lifting barbells? A psychologist might say that, unlike other creative endeavors, writing requires one to use both the "right" and "left" brains simultaneously. That is, the right brain, the source of creative power, must feed information to the left (analytic) side, which must, in turn, process the information into sentences that are not only intelligible, but also obey the rules of grammar. Or it just might be so hard because most of the time writers don't know *what* they want to say, let alone how to say it.

Perhaps, although knowing this doesn't make writing any easier. As long as we're being honest, we should also keep in mind that writing is like running, playing the piano, or any other rhythmic activity. That is, the hardest part is getting started. Once you start pounding the pavement or plunking the keys, you "get into" the activity and (amazingly enough) start to enjoy it. Writing is like that.

Speaking of this fascinating rhythm reminds me of an encounter I had with a fellow DuPont employee about 20 years ago. Early one morning, as I was rushing to get to my office, I literally bumped into my co-worker as he was running in the opposite direction. I didn't notice the sheaf of papers he was holding until I had knocked them onto the floor. I apologized at once and started helping him pick up the pile of typescript.

"Thanks," he muttered, and resumed his mad dash. "Hey, Terry!" I shouted. "Your office is the other way." "I know, but I have to get this report done before noon," he added breathlessly. "Can't work on it in the office. Too many distractions. Besides, I have to go somewhere I can get the flow."

Get the flow. Those three words speak volumes. Once you clear that first hurdle — call it "intuitive inertia" — the words do come quickly, if not easily. They start to tumble forth from your mind like a bag of marbles dropped onto a tile floor. At times, they spew out so fast that neither your pen nor even your word processor can keep up with them. That's when writing can be a veritable pleasure. And that's when you have the flow.

Alas, there are times when the flow is less like a raging river and more like sap dripping from a maple in January. No matter how hard you try, the words just won't come. Maybe you're tired, the room's too noisy, or your lunch wants to escape. Or maybe it's something else you just can't put your finger on. Does that mean you should quit and visit the coffee pot or comfort station? Possibly. Or perhaps you should just keep at it until the juices start to circulate once more. Only you can decide that — you and the deadline, that is.

Deadlines are synonymous with time; they are, at once, the bane and blessing of all writers. What writer has never uttered that woeful lament: "If only I had more time, I could've written a better report/article/ book. I was just too rushed to do a top job." Ah, well. If the writer had been told that he had, say, 2 days to write a 100-page report, this lament would've had merit. The fact is, most of the time we do have enough time to complete our writing projects. The real reason we feel pressed for time is that we've simply procrastinated until the last minute. No wonder we're biting the heads off our pencils — and everyone who comes around.

But how can time be a blessing? For the same reason it's a curse: it's just there. Without deadlines, we would have little incentive to write — or do anything at all, for that matter. Most of us are, basically, laid back.

(I know I am.) It's much easier to kick off our shoes, stretch out on the sofa, and read, watch TV, or snooze. Even those of us who enjoy writing, who take great pleasure in watching a pen glide across a sheet of paper, would probably rather be doing something else.

OF CLOCKS AND COLLABORATORS

This leads to the next point: writing is very time-consuming. Getting warmed up is only half the battle. The other half is to keep going long enough to finish the daily writing quota. How long does it take? That depends on the writer, of course. Some can generate pages upon pages of manuscript, ad infinitum. Thomas Wolfe, for instance, used the top of his refrigerator as a desk (he was 6'6") and would stand there for hours while he scribbled furiously, page after page, letting each finished sheet drop to the floor. After 8 or 10 hours, his feet were buried in a veritable avalanche of paper. Another Thomas, Thomas Mann, worked very slowly. An average daily output for him was a single page of longhand.

Most of us fall somewhere in between. I can't say whether this is typical, but it takes me about a half-hour to produce a draft page of double-spaced typescript (or two pages of manuscript, if I'm writing by hand). I need another half-hour or so to revise and retype it. In other words, each page requires about an hour of my time. The final typescript for this book will, I project, run to around 300 pages, including tables and figures. That's at least 300 hours, and it doesn't even include the time needed for researching, incorporating reviewer comments, and preparing the index. When you add that all up, the total time commitment is closer to 500 to 600 hours. These hours have to be carved out of my evenings and weekends.

Not every book requires this much time. In fact, many, especially the highly technical, require much more. A way to reduce the time burden is to enlist a co-author. He or she should be someone who is as knowledgable about the prospective subject matter as you are. Theoretically, the hours you'd spend writing would halve under such an arrangement. In reality, the reduction will be closer to one third, for a certain amount of time would be needed to communicate with your collaborator, to coordinate your research, plans, writing, and other tasks. (This assumes, of course, that the co-author will write his fair share and deliver it on time.)

The ultimate collaboration is a handbook to which each of 10, 20, or more authors contributes a chapter. If you happen to be the editor of this handbook, your work — writing work, that is — would be minimal, limited to editing the chapters, and perhaps contributing one yourself. The real work, however, would be in coordinating this massive effort and making sure each contributor knows what he/she has to write, how it should be written, and when it should be turned in.

But before all of that can be done, you have to enlist able and willing contributors — not an easy task, by any means. Some collaborators are extremely conscientious, while others are as reliable as an Icelandic weather forecast. While there are as many types of co-authors as there are people, collaborators tend to fall into certain categories:

- **Work Horse:** This kind of co-author makes an ideal team member. Not only does she complete her writing assignment on time, her work is invariably of the highest quality. Enough said.
- **Perfectionist:** This fellow is also a top-notch writer. However, he has one annoying flaw: a tendency to pick nits. He will submit not one, but several drafts of his assignment. Each draft will be better than the preceding, but the improvements therein will be invisible to the naked eye. Invariably, he will deliver his final draft (though this guy doesn't know the meaning of the word "final") on time. You can count on him to submit changes (probably by phone) long after deadline.
- **Procrastinator:** Though she may not exhibit annoying perfectionisms, this collaborator has no conception of time. She will begin writing her contribution on Saturday to meet a Monday deadline. As a result, her writing will look like a cut-and-paste job (which it is) and will be full of errors (which she will endeavor to correct only after considerable prodding on the part of the editor or principal author). However, once these errors are corrected, her contribution will be first-rate. (If only her work habits were half as good!)
- **Prima Donna:** This co-author has none of the faults of the above types. In fact, he's as good a writer as a "work horse" and has superior organizational skills. Unfortunately, he *knows* it. This is the sort of guy who regularly phones to ask how the other contributors are doing, as if to imply that they are not as efficient or as skilled as he. He will also ask such second-guessing questions as: "Do you think Dr. Larvae is the best contributor for that chapter on pine bark beetle eradication procedures? I know someone in our plant pathology department who can do a better job." In other words, he will try to usurp your authority as editor/principal author and do other things calculated to put you in a bad light and himself in a better one. You're well advised to keep this type at arm's length. Better yet, convince him to withdraw from the project, if you can. True, you'll lose a good contributor, but you'll also lose a pain in the posterior.

Regardless of the skills, reliability, or personalities of your co-authors, one problem remains to be solved. It is a problem inherent in many collaborative writing efforts: how to merge the different writing styles into a single, seamless "voice". Now, this problem usually doesn't arise with handbooks, as each chapter in a handbook typically is written by a different author. Nor does it arise if each author of a book (say, a textbook) is assigned certain portions and is given a byline for each of those. In these cases, there is no need to smooth over the stylistic differences.

But in other cases, where the respective styles of the collaborators must be melded into one, you, as editor or principal author, must strive to see that they do. This is by no means easy to accomplish. Even the best writers fail, despite their best efforts. A contemporary example is *The Talisman* (Viking/Penguin, 1984), a horror novel co-written by Stephen King and Peter Straub. This massive tome was by no means a critical success, and despite the drawing power of its authors (King, especially), had limited commercial success as well. There were several reasons why it failed. Among them were its story line, which meandered like an overfed bookworm throughout the 1100+ pages. But another reason was the collaboration itself. It just didn't work. King and Straub have vastly different writing styles. King's is chatty and colloquial, while Straub's is reserved and formal. Anyone familiar with their work can tell where the one leaves off and the other takes up.

Technical writing doesn't have to be as seamless as fiction. Yet it should be smooth enough to avoid jarring the reader to distraction. Making it so may require you to rewrite all or part of your collaborator's writing to make it resemble yours. But even if this isn't feasible, at the least you can redesign/redraw his tables/figures to make their formats identical to yours. You should do the same thing with his references and footnotes. Differences in these kinds of things are especially noticeable. The fewer the rough spots, the easier the read.

What if you have neither the time nor the inclination to write a book? How long would it take you to write, say, a journal article? Naturally, that would depend on the complexity of the material, the length of the article, the journal in which it is appearing, and the reason for writing it. The last has an important bearing on the others. Professionals write journal articles for various reasons: recognition, survival (especially in the academic world), money, and (in a few cases) simple enjoyment. And your reason for writing will influence the time you spend doing it. Odds are you wouldn't spend as much time writing an essay for a pop science-type magazine as you would a high-powered article for a refereed technical journal. Writing the former might take a day or two; the latter, weeks. The difference is, the refereed article requires more time for research, writing, rewriting, incorporating reviewer comments, and like chores.

"Time is money," of course, and you may very well decide not to spend any more time writing than you have to, especially if you won't be getting paid for it. That attitude might be justifiable *if* the article you're writing contains soon-to-be-dated material (e.g., cost data), *if* it's relatively nontechnical, or *if* it's the sort of piece that would not benefit from outside review. In such cases, the sooner you get the article into print, the better. If that means submitting to a nonrefereed journal, by all means do so. The peer review process can consume vast amounts of time.

However, if the material is complex, controversial, or profound, it pays to take the time to ensure that the article is the very best that you can write. How *much* time? Again, that depends on a number of factors. For talking purposes, we can develop rough time estimates, much as we did above for books. But like that "one hour-per-page" figure, these estimates are approximate, at best:

Type of Article	Time (hours/page)
Semitechnical	1.5–2
Moderately technical	2–3
Highly technical	3–4

Each time range denotes the hours required to produce a final page of article typescript (double-spaced) that would be suitable for submission to a journal. "Semitechnical" refers to articles that deal with technical subjects qualitatively, in a light-hearted manner, or both. The "You and Your Job" column in *Chemical Engineering* usually runs semitechnical pieces. One of the most entertaining YYJ contributions ever to appear was "The Chemical Engineer in the Kitchen," a tongue-in-cheek attempt to apply chemical engineering principles to the culinary arts. Another example of a semitechnical article was the "Mathematical Games" column that appeared monthly in *Scientific American* magazine. Written by the late Martin Gardner, the columns were a delightful mix of midlevel mathematics and real world applications of same, seasoned with a dash of dry wit. These kinds of articles are fun to write and generally require little research and rewriting.

Not so with the "moderately" and "highly" technical pieces. They still may be enjoyable to write, but they *must* be well founded and meticulously written. The difference between the "moderately" and "highly" technical is more one of degree than of kind. Moderately technical articles contain a fair amount of heavy reading — equations, tables, graphs, and footnotes — but usually enough expository material to permit them to be digested rather easily.

In highly technical articles, however, the equations, tables, figures, etc. are quite "heavy" and occupy well over one half of the allotted space. Moreover, the text (what there is of it) is typically turgid prose whose main purpose is, it seems, to provide a bridge between one set of data and the next. This is the sort of article preferred by refereed periodicals published by professional institutes and societies. I don't mean to be pejorative. Highly technical articles have their purpose, not only to advance academic careers but also to make lasting contributions to the field and to society at large. It's just that writing these articles demands a large investment of time and energy that only a few of us can afford.

By now, I hope that you've gotten "hooked" — hooked on wanting to learn about technical writing and what it can do to help you and your career. In the following chapters, we'll talk about how to start writing and how to get your writing published. Then we'll look at today's technical publishing market and expose ourselves to the legal and business aspects of the writing world. Finally, we'll learn what other technical writers have to say about their craft. But, before you turn the page, ponder what another famous writer, one Tennessee Williams, said about his art: "To write is to be *free*."

REFERENCES

1.   Van Gelder, R., *Writers and Writing* (New York: Charles Scribners Sons, 1946).
2.   Koontz, D. R., *How to Write Best Selling Fiction* (Cincinnati: Writer's Digest Books, 1981).

2   Some Preliminaries

Fifty percent of a writer's success has to do with perseverance.
Never give up. (The other fifty percent? Let's see...20% is talent,
20% is craftsmanship, and 10% is just pure luck.)
Dean R. Koontz, letter to the author.

In the last chapter, we discussed the pros and cons of technical writing and concluded that the pros predominate. If you've read this far, you probably agree with that conclusion. And you're no doubt ready to get down to the business of getting published. However, there is a lot of ground to cover between deciding to write and getting your name in print. In this chapter, we'll try to bridge that gap.

SPECIFYING YOUR SPECIALTY

Before he makes the first pen stroke, a writer needs to identify his area of expertise. This is easier said than done. Although we are told that this is the Age of Specialization, it is often hard to tell what our specialities are and, more to the point, how we can use our specialized knowledge to advance our careers.

Consider a federal bureaucrat who works for the U.S. Department of Labor as a labor relations specialist. Her job title and description tell little or nothing about her academic background. She could be a lawyer, a psychologist, or an engineer. Would this bureaucrat consider herself a lawyer/psychologist/engineer first and a labor relations specialist second? Or would it be vice versa? And if she wanted to get published to enhance her career, would she set her sights on a journal related to her academic background or one that dovetailed with her job specialty? Only she can make that decision. But at some point, she's going to have to decide which fork in the road to take. Such occupational identity crises are not uncommon in government, as well as in other large, highly structured organizations.

How do you identify your specialty? To start with, compare the work that you do to the work of others in your organization and others who perform the same or very similar tasks. (Notice that I didn't say "others with the same job title," as titles can deceive.) Are there common threads that run through your work and theirs? If so, your work is a bona fide specialty, whether or not personnel experts would have a classification for it. And if it is a distinct specialty, others in that specialty should have an interest in what you have to write about. Furthermore, there's a good chance that there's a periodical to provide an outlet for your writing.

But what if you can't locate any cohorts? What if you're the only person on the face of the earth who does what you do? Even then, you can still write about your work, but you'll have to do so for an outlet that publishes material relating to another field, possibly related to your academic background.

For example, before the environmental movement hit its stride in the 1970s and 1980s, there were few publishing outlets for air, water, and solid waste pollution control specialists. But because most of these specialists (then few in number) were sanitary, civil, or chemical engineers by training, they typically wrote for engineering journals. To get their writing published, they emphasized the connection between their environmental work and their well-established engineering disciplines.

The medical profession provides another illustration. Until recently, there were, essentially, two kinds of doctors: general practitioners (GPs) and specialists. The latter often looked down on the former with thinly disguised contempt. Consequently, the ranks of GPs thinned, as medical students took up the higher-prestige (and much higher-paying) specialties.

Then, about 20 years ago, the "family practitioner" (FP) was born. Now, in terms of who they treat and how they treat them, FPs and GPs are more or less the same. What distinguishes them is the fact that family practice has become a board-certified "specialty" with its own certifying authority, the American Academy of Family Practitioners. Now FPs (and those few GPs still practicing) have an outlet for their writing, *American Family Physician,* whereas before they had to slant their papers and articles toward the publication requirements of the *Journal of the American Medical Association* or *The New England Journal of Medicine.*

The point is, if you wait long enough, your esoteric job tasks may evolve into a bona fide specialty with its own publishing organ. Meanwhile, there are dozens of journals in which you can get your material published — if you can slant it in the right direction.

Another thing to consider is your interest in your job. Put simply, do you enjoy your work? If not, if you just consider it a means to a paycheck, odds are you won't have the interest or energy to get published in the first place. And even if you should find the spark to ignite your ambition, your lack of enthusiasm would show through in your writing.

If you do enjoy your work, however, your interest in it will continue to grow, and you will look for ways to improve your productivity and to make your work more meaningful to you and your organization. Moreover, you will see technical publication as a means to achieving that end, for publication will give you an outlet for your creative energies, as well as a showcase for your expertise.

WHAT DO YOU HAVE TO SAY?

After identifying your specialty, the next step is deciding what to write about. Would information about some aspect of your work be both interesting and useful to others in your specialty? Has anyone else written about this topic before? If so, who wrote about it and where was it published? Can you provide an angle on this information that would make it unique? These are just a few of the questions you should ask yourself before starting to write.

This reminds me of the first paper I ever wrote, back in the early 1970s. I had discovered (so I thought) a unique way to calculate the overall removal efficiency of several air pollution control devices linked together in series. The calculation procedure was based on rigorous mathematics, which is why it intrigued me so. Of course, I naively assumed that everyone else would find the procedure equally intriguing. I submitted the paper to a large-circulation engineering journal, one whose readership was (and still is) primarily practicing, as opposed to academic or research, engineers. To make a long story short, the paper was rejected. In his rejection letter, the editor wrote that the article was too theoretical for his magazine, specifically because the calculation procedure contained too many "undefined constants" for readers to make use of it. I sent it to a similar magazine and received the same sort of "thanks, but it's too theoretical" response. Finally, I gave up and had it published as an E.P.A. report.

A few months later, a chemical engineering professor (and part-time E.P.A. consultant) contacted me to ask if I'd like to revise my article into a paper and present it at a scientific conference. This I did. Shortly after the conference, the professor got my permission to submit the paper to a refereed (and well-respected) environmental journal whose editor he knew. I was delighted, for at last I was going to get my article published.

However, a few weeks later the professor phoned to give me some bad news: the journal had rejected the paper. "Why?" I asked, totally perplexed. "Did they think it was too theoretical?" "Not at all," he replied. "One of the peer reviewers told the editor that he had just read an E.P.A. report *on exactly the same subject!*"

So, in a way, I had plagiarized myself. This leads to another point; call it the First Law of Technical Writing: "Never reinvent the wheel — unless you can cover it with a new coat of paint." That is, don't write an article on a subject that already has been thoroughly covered, *unless* you are able to embellish it, give it a new slant, or otherwise make a worthwhile addition to that subject. And the only way to find out if indeed you have or have not reinvented the wheel is the hard way: **research**.

TESTING THE WATERS

Now, to many of us, research is an unpleasant chore. It means hours of digging through past issues of technical journals and reports, of scanning acres of microfiche, hoping to find (or, perhaps, *not* find) information on the subject we're planning to write about. Sometimes we're successful; sometimes not. But regardless of the outcome, we are usually left with the nagging feeling that we haven't checked every possible source, that somewhere out there is an article or report that we've missed. There's little we can do about that feeling, except to let it go. Because we must realize that, given the information explosion of recent years, it is virtually impossible to cover every conceivable source of information on a given subject. (We'll talk more about research techniques later in this chapter.)

Suppose that, in the process of digging, you have uncovered an article that says, amazingly enough, exactly what you want to say. Does that mean you should scrap your idea? Not necessarily. For one thing, just because you've found such a piece does not mean that the subject has been exhaustively covered in the literature. You may have found the only other article of this kind in existence. Or the article may have been written so long ago that it was no longer current, or the subject needed a fresh treatment.

About 5 years ago, I wrote a historical article on a facet of the Civil War. After doing extensive research, I became convinced that no one else had *ever* written about these obscure, yet historically important, events. After completing my research, I wrote the piece and submitted it to a popular history magazine. While waiting to hear from the magazine, I stopped by the library one day to research another subject. While poring over some papers in the manuscript department, I happened upon a clipping of a magazine piece on (you guessed it) exactly the same subject covered by my article. Was I shocked? Yes. Did I immediately phone the history magazine and ask them to return my manuscript? No, because the article had appeared over *60 years before.* Even if the piece had been published more recently, I wouldn't have been alarmed, because my article was so much different — in length, slant, style, and several other respects — that it was a totally new and original work.

And even if lightning should strike, and you should find a piece that is, for all intents and purposes, identical to the one you've planned to write, do not despair. You can still use it as a foundation for your article. Isn't that plagiarism? No. As we'll see in Chapter 6, plagiarism has a very precise meaning, etymologically and legally. According to Webster's dictionary, plagiarism is the "stealing and passing off the ideas or words of another as one's own *without crediting the source.*" (Emphasis supplied.) Webster's gives another definition: "presenting as new and original an idea or product derived from an existing source." Clearly, you would not be plagiarizing if you used another's work as the basis for your own, and if you made it very clear that you were doing so. In fact, scholars do this as a matter of course. A journal article typically begins with a section entitled "Prior Work" or "Previous Research". Without such a background, the paper would have less credibility, as the reader could assume (perhaps rightfully) that the writer hadn't done his homework.

CASTING YOUR BYLINE

Once you've decided what you're going to write — and are determined to make it original — you next have to decide where to place it. The "where", however, will depend on the "what" (i.e., whether you'll be writing an article or a book). Sometimes the dividing line between the two isn't very sharp. Length alone is no definitive criterion.

If length doesn't make the difference between a book and an article, what does? The subject matter does, for one thing. Typically, an article treats, in depth, a narrowly delineated subject, while the scope of a book is much broader. A book covers several topics, each of them distinct but all related, much like the members of a large family gathered together for a special occasion. Another analogy can be found in the world of fiction. Here, the distinction between short and long forms is relatively clear-cut. Traditionally, a short story portrays a single incident via characters that are minimally developed. A novel, on the other hand, consists of (or at least should consist of) a succession of incidents involving fully dimensioned characters that, acting together, work toward the furtherance of a distinct theme. That is, a novel is to a book as a short story is to an article — more or less.

Most of the time, though, the subject matter will dictate the form (article or book) and the medium (periodical or book publisher). Few articles are broad enough to be expanded into books, and few books are so narrow in scope that they could be effectively synopsized in one or even a series of articles.

Before going further, we need to consider what an article is. Actually, it can be several things: a "special report", a feature piece, a technical note, a book review, even a letter to the editor. A writer can use each of these forms to convey technical information, according to its unique characteristics. For example, a technical book review must list the book's title, author, length, and list price, as well as the publisher's name and address. A review also should describe the book's contents and evaluate their accuracy, completeness, clarity, etc. However, a reviewer can also use the review to convey *her* knowledge about the subject matter covered in the book. (In fact, the reviewer would not be reviewing the book in the first place, unless she had at least a working knowledge of the subject.) The reviewer can present experimental results and use these to confirm or refute the book's conclusions about a certain point. She can also compare the author's ideas to those of other experts in the field with whose work she is familiar.

Articles (reviews, features, technical notes, etc.) are published in periodicals, variously titled "magazines", "journals", "transactions," etc. Although their formats, contents, and editorial policies vary considerably, most technical periodicals have certain common features: (1) periodic publication, (2) acceptance of contributions from outsiders ("freelancers"), and (3) editorial staff and policy. For some periodicals, "periodic" means quarterly, while others appear bimonthly, monthly, or weekly. The frequency of publication depends on several factors, such as the timeliness of the material, the number of contributions, and the readership.

Nearly all technical periodicals publish articles submitted by freelancers. It would be difficult, if not impossible, for the editors to find the time and energy to research and write technical articles on the range of subjects covered by the magazine. And even if they somehow could, such a let's-keep-it-in-the-family editorial policy would deprive readers of the exposure to a variety of ideas and concepts. Also, most periodicals give each contributor a byline and include his biographical sketch and, in some cases, photo. For an author interested in getting peer recognition, such things are essential.

Periodicals can be divided into two categories: "refereed" and non-refereed. I've used the former term before now but haven't explained what it means. "Refereed" or "peer-reviewed" periodicals are those that will not accept a manuscript for publication unless it has been cleared by two or more reviewers, preferably those with expertise corresponding to the manuscript subject matter. These are the kinds of periodicals that academic and research professionals usually prefer because acceptance of their work in refereed journals gives impetus to their careers. The longer one's list of peer-reviewed publications — especially in the "top" periodicals — the higher one's professional standing. This is especially true if one is the sole or principle author of these publications. (In fact, conventional wisdom has it that one's career standing is no longer measured by the number of refereed publications, but by where and how often one's publications have been *cited by other authors.*)

At any rate, the acceptance procedure for most refereed periodicals is rather involved and time consuming. First, you would submit to the editor a manuscript that has been carefully prepared to the periodical's guidelines. Many of these guidelines are quite exacting. Consider the following excerpt from the "Instruction to Authors" for *The Engineering Economist* published by the Institute of Industrial Engineers (25 Technology Park, Norcross, Georgia 30092):

> Manuscripts should be typewritten, double spaced, with wide margins. Mathematical notation for interest factors should follow the functional format of ANSI Standard Z94.5-1982…Authors should attempt to make mathematical expressions in the body of the text as simple as possible. If equations are numbered, the numbers should be given in parenthesis, flush with the right margin of the page. Authors should restrict themselves to the English alphabet and standard Greek symbols when choosing symbols. Subscripts or superscripts should be clearly below or above the line. Lengthy mathematical derivations should be placed in an Appendix…

Next, the editor sends your manuscript to two or more peer reviewers with whom he is acquainted but whose names he would rarely (if ever) divulge. He asks them to review the manuscript and to return it, with marginal comments, along with a completed form. On this form, the reviewer will indicate such things as whether the manuscript should be published as is, published with minor changes, published after a major rewrite, or whether it should be rejected altogether. (See Figure 2.1.) Once the editor receives these forms and the other reviewers' comments, he usually forwards them to you. Depending on the nature of the comments (and the editor), you will be asked to incorporate some, all, or none of the comments into the manuscript. Of course, if most of the reviewers recommend rejection, a rejection letter is what you will get. But these are not hard and fast rules. If for example, you have published frequently for this periodical and/or have a well-established reputation in your field, the editor might simply tell you to do what you would with the comments — incorporate them or dispose of them. That is, for a few authors, the peer review process is merely a formality designed to "test the waters" before the article is published. However, for most writers, the process has much more substance. Though painful in some instances and ego shattering in others, the

Figure 2.1. Typical peer review form.

review process invariably provides very useful feedback that would enhance the quality of nearly any manuscript.

Finally, once you have revised your manuscript in accordance with the comments, you send it back to the editor who then schedules the article for publication, typically 3 to 6 months later. The published article usually bears an uncanny resemblance to the final manuscript, both in form and content. This is because, at refereed journals, substantive and even grammatical editing is often minimal. (Hence, the exacting manuscript preparation guidelines discussed above.)

With non-refereed publications, the entire process is more straightforward. You prepare your manuscript, usually to less demanding specifications than those of refereed periodicals and send it to the magazine. The editor then circulates the manuscript among the editorial staff to obtain their critiques. If the majority vote "yes", the manuscript is accepted. If not, a rejection letter follows. The latter usually contains some explanation of why the piece was rejected. (In this respect, the letter functions like the review forms provided by most refereed periodicals.)

If the article is accepted, do you ever have to revise it before publication? Not very often, because the editors have accepted it on its own merit, which is another way of saying that they trust your technical judgment and expertise. "Then," you might reason, "the published article will be a carbon copy of the manuscript. Correct?" No. At non-refereed periodicals, manuscripts are usually edited, sometimes even rewritten, to suit the periodical's length and style requirements. Tables and figures might be redrawn by graphic artists. Photographs might be added to spruce up and provide breaks in the text. In all, the editors take a more active involvement in the manuscript preparation than at refereed periodicals where the author is usually expected to provide the entire package — writing, editing, and graphics. This does not mean that the author is never involved in this process. On the contrary, the editor will usually contact you if he wishes to make substantive changes to your manuscript. In addition, before the advent of computerized typesetting, the editor would usually send you "galleys" of the edited manuscript for you to review. In this way, editors function much as they do at "trade" (mass market) periodicals. (We'll talk more about dealing with periodicals in Chapter 3.)

BETWEEN THE COVERS (TYPES OF TECHNICAL BOOKS)

Suppose your writing won't fit the confines of an article. If that's so, then you'll have to start thinking "book". Now, just the mere mention of that word gives pause to many professionals. They envision spending months slaving away to produce a manuscript and months more making corrections to it, reviewing proofs, writing the index, and all the other tasks, big and small, that comprise the book production process. It cannot be denied: a commitment to write a book is indeed a major one. But, sometimes, a book is the only vehicle available to convey the information you need to impart. We'll explore the world of book publishing in Chapter 4. For now, let's just consider the categories of books that your writing might fit into.

- **General technical:** This category includes the majority of technical books published. "GTBs" can (and do) deal with just about any subject, from agronomy to zoology. Although they are often written with a particular audience in mind, their subject matter is accessible to a wide range of readers, some of whom may not even work in the field.

Many technical books contain examples to illustrate applications of the subject matter. These may be anecdotes, sample problems, case studies, or all three. Typically, GTBs also contain a large number of illustrations (drawings, photographs, etc.), tables, and figures. Appendices may also be included to cover material that is too detailed for the main text. Finally, a subject and, occasionally, a name index is provided. However, GTBs are usually *not* suitable as textbooks, mainly because the subject matter is not pedantic enough and because the examples are too few to provide enough analytical exercise to students. More practically, their bindings and paper stock are usually not sturdy enough to withstand the wear and tear of the classroom. On the other hand, GTBs can be, and often are, used for supplementary reading, especially in graduate schools. So if you intend to write a technical book, keep the foregoing points in mind.

- **Textbooks:** Where GTBs have a (relatively) broad readership, texts are dedicated to a specific discipline or even a specific course. Often, a college instructor will design a course to fit a particular need, collect materials for this course, and then compile the materials into a textbook for use in his classes. This doesn't prevent a nonacademic from writing a text "around his material," especially if the material is rigorous, highly structured, and dovetails with an established curriculum. For this reason, textbooks are more difficult to write and require considerably more peer review. And after all this, they can be difficult to sell, mainly because they have to compete with books written by academics who, by virtue of their background and insider status, have the upper hand.

- **References:** In some respects, reference books represent an untapped market for writers. Well-known references, such as the *PDR (Physicians' Desk Reference)* and the *CRC Handbook of Chemistry and Physics*, seemingly dominate their respective markets in that they give both readers and writers alike the impression that between their covers lies all the world's knowledge on their subjects. Hence, such books are very popular — and very expensive. Moreover, they are typically edited, not written, as the information is so broad and complex that no single person could possibly compile it all.

  This doesn't mean that a writer should shy away from writing reference books. If your subject matter is narrow enough to grasp, yet broad enough to fit a reference or handbook format, then by all means consider this form. An immensely successful handbook, one that has sold millions of copies, is Strunk and White's *Elements of Style*. This excellent little book clearly and consisely presents the basics of composition for all writers, not just the technically inclined. Yet, if you examine it closely, you'll see that it's structured like a handbook, not a GTB or text. A reader can easily find what he wants to know about a given topic, be it punctuation, syntax, or style.

Perhaps your handbook material is too broad to cover alone. For example, you may be an accomplished general surgeon and wish to write a book on modern surgical techniques. But, you do not think that you know enough about surgical specialities, such as neurosurgery, to write intelligently about them. In such a case, you should consider enlisting a corps of surgical specialists. Your co-authors would fill the "knowledge gap" and, due to their professional reputations, would likely add credibility to your book, which undoubtedly would be classed as a reference. However, as we discussed in Chapter 1, working with co-authors requires management skills, editing ability, powers of persuasion, and above all, patience (all this in addition to writing ability).

Nonetheless, the efforts of hard-working editors and contributors have produced such classic references as *Perry's Chemical Engineers' Handbook.* One might wonder if the work required to edit one of these tomes is worth it. Only their editors can say. But consider: Perry's Handbook has gone through six editions and dozens of printings since the first edition appeared in 1934. Surely, it has earned the editors — John Perry, then Cecil Chilton, and most recently Don Green — a tidy sum. More importantly, it has earned them solid and secure reputations in their field.

If a "Perry's" would be too ambitious an undertaking for you, a smaller, more specialized reference book could be what suits you and your material. Suppose that during months of intensive work your research team has collected reams of data on the speed at which white, black, and spotted rats require to negotiate a variety of mazes. Those data might provide enough raw material for a journal article, but not enough for a book, let alone a handbook. Suppose further that your team has also measured the heart rates and blood pressure of said rodents as they ran through these mazes. And what if they also had observed their biological functions before, during, and after these activities? Finally, you and your colleagues also may have observed the rats' physical changes (e.g., body weights) and social development associated with these labyrinthine experiments. You *could* compile all of this in a technical book, but the range of material covered would be very broad and not too deep in any one place. However, a small handbook might be just the ticket, perhaps something entitled Our A-MAZE-ING Rats: A Compendium of Physiological Observations. Inasmuch as you and your team have collected the material, you can proceed to write it up on your own (after giving your colleagues due credit, of course).

EXCAVATING (LIBRARIES AND OTHER INFORMATION SOURCES)

Books are to articles what cats are to dogs; they are related, they require a lot of work and T.L.C., but they are so inherently different that they are essentially incompatible. The approach the prospective writer would take when undertaking a book would be inherently different from that taken for writing an article, and vice versa.

But what books and articles do have in common, apart from the commitment of time and energy, is the need for *thorough research.* Research not only helps you get where you're going, but also tells you where other writers have been. Thus, the first step in any research program is to read and study everything that relates to the subject you're planning to write about. For some topics, that can mean a few books and articles; for others, entire libraries. Of course, few of us have the time or stamina to devour shelves of books on a given subject. But, the massiveness of a body of material should not deter you from selecting some parts of it. The most time-effective approach in such cases is to obtain the most recent works on the subject, those published within the past 5 years. Anything older than that would likely be outdated and of little use, save for providing a historical perspective. Why 5 years? Apart from being a convenient and easily remembered span, 5 years is typically the time required for one to (1) perform research, (2) collect and compile data, (3) obtain a publisher, (4) write a book, (5) get it published, and most of all, (6) have the book read and accepted by one's peers. In some cases, it takes less time, but mostly it takes this long or more.

With this 5-year rule of thumb in mind, begin your research. Your first stop shouldn't be the library, but your own office. That's correct. Some of the best research materials can be found in your own files. No doubt you've been keeping files of journal clippings, company reports and memoranda, and other items of interest and possible use. Start poring through these files; you may be surprised at what you'll find. When you're finished with your cache, ask to look at your co-workers' files. Their files could contain just the material you've been looking for.

Your company's library should be your next stop. Many firms (including consultants) maintain in-house libraries for the benefit of their employees. The larger of these are staffed by professional librarians whose duties include purchasing books, reports, and subscriptions to periodicals that pertain to the firm's products and services. More importantly, these libraries collect and catalog company reports, related to research, marketing, sales, and other topics, which can be found nowhere else.

Next, visit the nearest outside library. Which library (or libraries) you visit naturally will depend on your subject matter. For some subjects, such as law or medicine, the choice is easy. But what if your topic is rather esoteric, for instance, the manufacturing and application of exterior latex paints? Should you visit a university library devoted to chemistry and other sciences? Probably, but also stop by the local library, for that might contain less technical (but equally important) information on related subjects (e.g., home remodeling).

Since every library offers different kinds and levels of services, it is impossible to say what would be available to you there. However, based on discussions with university and public librarians, I have compiled lists of tools that, if properly used, can build the foundation of any research project.

The first list is typical of a public library serving a medium-sized urban area.[1]

- **Books:** over 310,000 fiction and nonfiction titles, including two special reference collections and many large-print books.
- **Audio-visual services:** albums, cassette tapes, video tapes, compact discs, 16-mm films, and books on tape, all circulating.
- **Public Access Catalog (PAC):** a computerized, interactive system that offers quick access to information about materials (books and audio-visual items) in the library system, including the location and status of each copy.
- **Reference Services:** a collection of materials grouped into several specialized services and collections:
  1. Business reference — provides U.S. and international company directories, investment sources, tax guides and forms, annual reports, periodicals, telephone directories, etc.
  2. Community information — provides information on community agencies and services
  3. Interlibrary loan — means to access materials in participating libraries across the U.S.
  4. Periodicals and microforms collection — provides more than 800 magazines and newspapers and a variety of index services (e.g., "InfoTrac II")
  5. "Project LIFT (Learning Information for Today)" — offers educational and career information, job listings, literary materials, etc.; also offers resume and job-hunting consultations

A university library would offer similar services, though theirs would, of course, be geared toward the academic community rather than to the public at large. Some of the services unique to a large private university library are:[2]

- **Reference:** reference librarians are available to answer informational questions, advise library users about research strategy and services, conduct searches of automated data bases, compile bibliographies of library materials, and provide instruction in library use.
- **Public documents and maps:** includes government and international publications, as well as topographic, geologic, and special subject maps.
- **Newspapers, microforms, and periodicals:** both popular and scholarly periodicals are offered, in both "hard copy" and microform; equipment for reading the latter are also provided.
- **Interlibrary loan:** as described above, this allows the library to obtain materials housed in other locations. In this library, an entire department has been dedicated to handle these loans.

- **Online catalog:** allows a user to access, via computer terminal, information on any of the library's extensive holdings, as well as those housed at other area university libraries; the catalog may be accessed via the library's terminals or any personal computer equipped with a modem.
- **Special collections:** housed in a separate department, these consist of unique, scarce, and valuable materials, principally manuscripts and rare books, that must be secured in a controlled environment; examples of the former are letters, diaries, account books, printed materials, and photographs.

Finally, you may need to consult a specialty library. Most of these are located at universities and are, in fact, satellites of the main library. Although there may be some overlap between them, the specialty library will house books, periodicals, and other materials related to that field that would be too esoteric for the main library. For example, the engineering school at the university whose library was profiled above "contains a working collection of publications in both computer science and engineering".[2] To give some idea of the extent of these holdings, this library alone subscribes to over 800 periodicals.

This and the other libraries at this university are very accommodating to the general public. They issue free library cards, provide free research services, and otherwise serve the "town" as well as the "gown". This is *not* the norm with all universities, however. In other cases, those not affiliated with the university may be given only restricted library privileges, if any.

Lastly, if your files, your firm's library, and your local public and university libraries don't have what you need, your Uncle Sam just might. The three branches of the U.S. Government publish thousands of reports every year. Some of these reports can be obtained directly from the governmental agencies that prepared them, often at no charge. However, you have to know who to call and the title and number of the report. That may sound easy, but it's not. A federal agency is a complex labyrinth of offices, bureaus, divisions, and other entities, the names of which often bear little or no resemblance to the kind of work they do. Even if a researcher had a detailed organizational chart of an agency, he might be hard pressed to locate the group that produced a report on a given topic. If he was persistent (and lucky), he might actually locate the report's author or project officer (if the report was prepared by a contractor); but, the author might just tell him something like, "We released that report 3 months ago, but we're out of copies. You'll have to get it from NTIS. Here's the report number..."

NTIS stands for the National Technical Information Service. This is a huge clearing house of government documents headquartered in Springfield, Virginia, a Washington, D.C. suburb. Nearly every federal report released to the general public is available through NTIS. According to its *Products and Services Catalog,* NTIS can make available over 1.6 *million* technical reports, either as paper or microform copies.[3] These reports deal with a wide range of subjects, ranging from "administration and management" to "urban and regional technology and development". More than just a reports clearing house, NTIS provides many products and services useful to the researcher. For instance, NTIS...

- Announces 150,000 summaries of completed and ongoing U.S. and foreign government-sponsored R&D and engineering activities each year; these summaries detail the knowledge and technology generated through [the] U.S. Government's $60 billion R&D effort.
- Provides the complete technical reports for most of these results.
- Manages the Federal Computer Products Center, which provides access to software, data files, and databases produced by federal agencies.
- Manages the Center for the Utilization of Federal Technology (CUFT) which runs the most active inventions licensing program in the U.S. Government. CUFT also links U.S. firms to federal laboratory contacts and technologies with its directories and catalogs.
- Brings U.S. firms access to more than 15,000 foreign government research and engineering programs.

Moreover, NTIS is a unique government agency, in that its operations are supported solely by sales revenue. No congressional appropriations are made to cover these costs.

Their prices are not cheap, however. Microfiche and paper copies of reports range from $8 to $80 each, depending on the report "code" and whether its price is "standard" or an "exception". The Federal ComputerProducts Center prices run from $50 to $930 apiece for microcomputer diskettes, while magnetic tapes cost from $165 to $2,100 each. Finally, these prices apply only to customers in the U.S., Canada, and Mexico. For others, the prices are approximately twice the base rates. Furthermore, paper copies are almost always photocopies and, as such, may be of poor quality; because the copies are black and white, they do not pick up the shadings and tones evident in color illustrations.

If you know neither the federal report title nor its number, or if you just want to know what reports are available for a given topic, the NTIS publishes the *NTIS Title Index,* a "quarterly cumulation of titles for sale by NTIS...arranged for easy searching with a keyword, author, and report number index."[3] (Price: $375.)

With all the information available from sources, private and public, there is little likelihood that you'll be unable to find enough research materials for your writing. Whether you'll be able to find the time and energy to undertake this research, only you can say; but just as no structure is any stronger than its foundation, no article or book is any better than the research supporting it.

REFERENCES

1.   Durham County Library, Durham, NC: Durham County Library.
2.   *A Guide to Duke University Libraries* (Durham, NC: Duke University Publications Committee, July 1990).
3.   *1990 NTIS Products & Services Catalog* (Springfield, VA: U.S. Department of Commerce, National Technical Information Service, January 1990).

3 Getting Yourself Published: Part I — Articles

If you would not be forgotten as soon as you are dead, either write things worth reading or do things worth writing.
Benjamin Franklin

Given the large number of technical book publishers and technical periodicals in business today, it would seem that the market for your article or book would be that of a seller's, not a buyer's. It would seem that, as a writer, you would be in an enviable position. However, when you look closer, you'll see that the pickings are leaner than sheer numbers would indicate, especially for periodicals. The reason is, most periodicals publish material that falls into a relatively narrow category.

The same can be said of technical book publishers, though the scope of their interests is usually broader than that of the periodicals. Some prefer to publish books on medicine, while others specialize in environmental, business, scientific, or other titles; and one publisher does little else but reprint federal government reports, which cannot, by law, be copyrighted. Of course, some firms, particularly the larger ones, publish titles covering a range of topics. We'll cover today's technical writing market more thoroughly in Chapter 5. In this chapter and the next, I'll present tips and suggestions that, if used, will make it easier for you to get your article or book published. Since more technical professionals begin their publishing experience by writing articles than books, we'll begin with the journals. We'll deal with books in Chapter 4.

DECIDING WHAT TO WRITE AND FOR WHOM TO WRITE IT

Clearly, a writer must get informed about the market if he expects to maximize the impact and benefits of his work. This is especially true for periodicals. Unfortunately, the average technical person doesn't give this sort of "market research" much attention. He does his (technical) research, writes his article, then looks for someone to publish it — when, in fact, he should do the last thing first. Before writing a word, he should decide whether to submit his article to a refereed or a non-refereed periodical. (We discussed the features of both types in Chapter 2.) Next, he should contact several periodicals to see if his work would suit their requirements. Finally, he should tailor his writing as much as possible to the needs of one of these periodicals. This doesn't mean his trying to pound a square peg into a round hole. Rather, it means establishing a working relationship with the editor of the periodical, one that will both benefit the writing and enhance the odds of getting it published.

How and where does one find the right periodical? To answer that question, we first need to consider some "selection criteria" to help us shorten the list of journals. First, there is the *subject matter*. This may seem to go without saying, as only a foolish writer would consider sending his work to someone who'd have no interest in it. But, there's more to it than that. Two periodicals may regularly publish articles on a given subject, say chemical plant design. However, the first may prefer material that appeals to (and can be readily used by) industry engineers and scientists, while the other publishes pieces slanted toward the research and academic communities. The latter articles likely would be highly technical and might require more extensive peer review, while the former would be less technical and might require no peer review at all. Though the topics of these two periodicals cover may be alike, their subject matter is vastly different.

The next criterion is *type of article*. Do you want to write a feature article? shorter piece? book review? technical note? letter to the editor? Have at least a rough idea of what you want to write before you contact a periodical. For instance, if you like writing book reviews (a good way both to get published *and* collect free books!), target periodicals that run reviews. Not all of them do. (In fact, the number that do is shrinking.)

Targeted audience is the third criterion. If you want your work to be read by the academic/research communities, you would naturally slant it to the more technical journals. Your readers would probably have similar interests, academic backgrounds, possibly even job descriptions. You should keep these readers in mind when you write your article. In that way, you can achieve a closer relationship with them, thus making it easier to avoid such mistakes as writing down to your readers or writing over their heads; and if your writing can be tailored to your readers, they will give it a more favorable reception.

All well and good, but how can you establish this relationship? There are as many ways of doing this as there are writers. A technique that works for me is to mentally burrow inside a typical reader's head. This "typical" reader is an imaginary person whose background, interests, etc. reflect those I'm targeting. If those readers happen to be air pollution control engineers, for instance, I would put myself in the shoes of one of them for a day. I would be with him when he answers his incessantly ringing phone, attends daily and weekly progress meetings, evaluates permit applications, and goes on site visits. And then I would ask myself, "Would what I plan to write interest or help this engineer do his job? If so, how should I write it to get the information across clearly and concisely?" I know that this engineer is very busy and has little time for reading the daily newspaper, let alone technical periodicals. I have to somehow capture this reader's attention with my writing and keep him reading it until the end.

Circulation is the next criterion. A technical periodical's circulation is an important, easily obtained, but oft-ignored statistic. Other things being equal, the larger the periodical's circulation, the more people will read your article. A second axiom: the broader a magazine's subject matter, the bigger its circulation and, in turn, the greater the impact of its articles. Of course, if you are pursuing a very narrow line of research in an esoteric field, circulation may not be important, especially if only a very few periodicals would publish your material anyway. For you, the primary objective is to get your work into print, regardless of how many people would read it. But if only a small number have the opportunity (let alone the time) to read your article, how can you expect to build a reputation in your field? Your article might be like the proverbial tree in the forest whose falling is heard by no one.

A fifth consideration is *readership*, as opposed to circulation. All periodicals have what is known as a "pass along ratio" (P.A. ratio); that is, every copy purchased or distributed to subscribers is passed to others who read it. The higher this ratio, the higher the readership. There is probably some relationship between the magazine's subscription price and its P.A. ratio. (That is, the more it costs, the more it's likely to be passed along.) The proliferation of photocopiers also has some effect. These days, it's quite common for a subscriber to photocopy an article and distribute copies to co-workers, copyright laws notwithstanding. Regardless of whether the entire magazine or photocopied portions is passed along, the increased readership cannot fail to help writers. These ratios vary a great deal, from less than 2 to 5, 6, or more. To learn a periodical's P.A. ratio, ask the editor-in-chief.

The final, and perhaps most important, periodical selection criterion is *editorial policy*. This covers a host of things, from manuscript format to honoraria to copyrights to complimentary copies. Some of the policy (e.g., manuscript preparation format) may be gleaned from reading back issues of the periodical; but most must be obtained directly from the editor or publisher. Many periodicals publish contributers' guidelines which address these and related matters which are available upon request. If the guidelines don't answer all of your questions, write the editor-in-chief a letter like the following:

> Dear Mr./Ms._____:
>
> Thank you for sending me a copy of "Contributor's Checklist for *Journal for Advanced Cybernetics*". It is very clear and informative. However, the checklist does not address certain important facts, such as (1) the time you need to review an article proposal, (2) the time between acceptance of a manuscript and publication in the journal, (3) the number of article reprints you provide contributors, and (4) your current readership (including pass alongs).
>
> Very soon, I expect to send you a proposal for an article related to my research program. Consequently, I would appreciate your sending me this information as soon as possible. A S.A.S.E. [self-addressed, stamped envelope] is enclosed for your convenience. Thank you.
>
> Sincerely yours,
>
> I.M. Sangfroid, Ph.D.

Notice that this letter also contains a question regarding readership. There is no law saying that your inquiry should be solely confined to editorial policy matters. In fact, the more information that you can get from a periodical before submitting a proposal, the better.

Before we go any further, let's summarize our periodical selection criteria:

- Subject matter
- Type of article
- Targeted audience
- Circulation
- Readership
- Editorial policy

By the time you've exercised these criteria, you'll have probably identified at least several possible periodicals to contact. You've also likely obtained basic information about them — title, address, phone number, editor's name, etc. If you haven't, however, there are certain references that can help you collect these data. The most comprehensive of these references for periodicals is *Ulrich's International Periodicals Directory* which lists over 100,000 periodicals. (We'll discuss it and other references more extensively in Chapter 5.)

GETTING YOUR FOOT IN THE DOOR

Deciding who to send proposals to is only the first step; doing it effectively is something else again. First, realize that, like you, magazine and book editors are busy people. Their work days are crowded with meetings, phone calls, reading and editing manuscripts, writing letters, and other tasks. Hence, the time that they can devote to reading proposals is limited, perhaps to a few hours per week. So, if you want your proposal to be given full consideration, make sure that it is clear and concise. In this section, we'll show how to prepare a proposal that not only gets and keeps an editor's attention, but which also improves the odds of your writing being accepted.

Let's define "proposal", or "query", as it's often called. Used here, "proposal" refers to written information that attempts to sell a piece of writing that is completed, partly completed, or yet to be written. A proposal may be anything from a one-page letter to a multi-page outline. We'll look at these and other kinds of proposals in turn.

QUERY LETTERS

A query letter is the briefest proposal, but is often the most difficult to write. Given an editor's busy schedule, a query letter should be as short as possible, preferably one page or less. In this short space, you must (1) summarize your proposed article, (2) explain why it is different from (or better than) other publications on the subject, and (3) tell something about your professional and/or writing experience. Moreover, your letter should avoid unnecessary detail, be written in crisp, clear language, and most of all, grab the editor's attention and keep it.

Before beginning the letter, take the time to learn the editor's full name and title. This is as important as finding out her mailing address. Letters that begin with "Dear Editor" or worse, "Dear Sir or Madam" get off on the wrong foot and tell the editor that you were too lazy to obtain this vital information.

The next step is the most difficult: grabbing the editor's attention. A query letter that begins with, "During the past 3 years, our research team has collected data on the heat generated by various sizes of fluorescent light fixtures, so as to determine the effect of this heat on room temperature and comfort..." won't get very far. However, this opening might: "Are your office lights making you sweat?" Now that you have her attention, you can go on to write the rest of your query letter, covering the topics listed above. (See Figure 3.1.)

At this point, you may ask, "I'm a scientist (physician, engineer, lawyer, etc.), not a snake oil salesman. If I write something like this, I'll look like a fool." No, you won't. Everyone has a sense of humor, editors

1947 Roe Lane
Guttenberg, PA 16147
February 29, 1991

Mr. D. Leete
Managing Editor
Ergonomics Today
333 Open Space Boulevard
Faraway, TX 78987

Dear Mr. Leete:

Are your office lights making you sweat? Or are they causing you to squint as you review manuscripts or queries like this one? The fact is, Mr. Leete, that the fluorescent fixtures in most American offices are causing workers problems that, at least, make them uncomfortable; at most, very sick. Squinting and perspiring are just two of the lesser hazards. Others include hearing loss (due to the buzzing ballast units) and excess sigma ray doses. As you know, the latter can cause metronoma, an especially pernicious cancer of the earlobes.

I would like to make the readers of *Ergonomics Today* aware of the health problems that their office lights pose. I propose to write an article for your "Work Space" department. This article would be written for *all* office workers, from receptionists to corporate executives — especially the latter. Although every office worker deserves to know about the dangers of fluorescent lighting, only managers have both the resources and authority to identify and correct the problems it creates.

My article would build upon the landmark workplace research reported in the literature since 1985. This includes Professor Cubicle's excellent paper on the hazards of shattered light bulbs and Dr. Modular's monograph, *Indoor Illumination: An Annotated Bibliography*. These and other studies — including your magazine's recent series on the relationship between worker productivity and office window area — should provide an excellent foundation for my proposed article.

As an ergonomics engineering consultant, I'm intimately familiar with the office environment and the hazards it can present. I earned both my B.S. and M.S. in Industrial Engineering from Wheatland (PA) Technical Institute, and have 22 years of industrial and governmental experience in my field. During the past three years, I've published one book (*Optimizing Your Space,* Lewis Publishers: 1989) and seven articles on ergonomics and related subjects for such periodicals as *Desk and Chair* and *Ceiling and Four Walls*.

If you have any questions or would like more information, please write or phone me at (777)-555-3333, from 9 to 5, Monday to Friday. For your reply, I have enclosed a self-addressed, stamped envelope. I look forward to hearing from you soon.

Thank you very much for considering my query.

Sincerely yours,

Harmon E. Interior, P.E.

Enclosure

Figure 3.1. Sample query letter.

especially. Besides, after reading pages of technical material all day, a query letter that begins with such a brassy statement is bound to relieve some of her tension, and probably evoke a chuckle or two. At the same time, a sentence like this will tell the editor that you have a sense of humor, that you *don't* take yourself too seriously, and finally, that you're probably going to be easy to work with.

The summary of your article comes next. A brief paragraph is all you'll need. Tell what you've written (or plan to write), why, and for whom. Avoid generalities like, "This article should be helpful to engineers." Instead, write: "The organic chemical waste handling techniques described in this article should educate plant engineers responsible for environmental quality maintenance at their facilities."

In the next paragraph, mention other pieces on this subject that you've read, *especially those that have appeared in the periodical you are querying.* This not only tells the editor that you've done your homework, but also says that you're familiar with the periodical. Also, explain how your writing will augment the body of knowledge on this subject. Or, if it's on a topic already well covered, explain how yours is better. Don't be modest either. Your objective is to get your work into print and, thereby, help many others as well as yourself.

The third paragraph of the query letter functions as a mini-resume. In a few sentences, tell about your professional training, your present position, and your previous publications (if any) in your field. Don't list your publications, however. Instead, say something like: "During the past 3 years, I've contributed three articles to *Plant Pathologist* and two to *Deciduous Digest*." The reference to "3 years" merely tells the editor that your publications are fairly current.

Close the letter with a paragraph that begins something like: "If you have any questions or need additional information, please write or phone me at 777-555-3333, from 9 to 5, Monday to Friday." Add that you have enclosed a S.A.S.E. [self-addressed, stamped envelope] and, finally, thank her for considering your proposal. Figure 3.1 contains each of these elements.

GETTING A SECOND OPINION (THE PEER REVIEW PROCESS)

Few periodicals are lone ranger operations, at least in editorial matters; that is, an editor rarely decides to accept or reject an article proposal without first getting additional input. In most cases, he circulates a proposal among the staff to obtain their opinions of it. Alternatively, he may send it to outside ("peer") reviewers. These reviewers (who are sometimes compensated for their services, incidentally) evaluate the proposal (or, if appropriate, the entire article) and return their comments to the editor. For everyone's convenience, the reviewers often jot their comments on a review form supplied by the periodical/publisher. (For a sample review form, see "Typical peer review form," Figure 2.1.) Based on these outside and/or in-house comments, the editor will decide whether to accept or reject the proposal/article. In Chapter 2, we discussed how the peer review process works; you may want to reread those portions now.

SEALING THE BARGAIN

At last, the day you've long awaited is here; the periodical has notified you that your proposal has been accepted. You'll get the go-ahead to write your article, perhaps along with a few suggestions regarding length, slant, and deadline. On the other hand, if it's a complete article, the editor will probably send along his or the peer reviewers' comments, as appropriate.

Before writing your article or revising one you've already written, write the editor a short letter acknowledging his acceptance and his comments. If you haven't already, this is a good time to discuss such matters as deadline, payment/honorarium (if any), publication date, etc. From the author's standpoint, it's always much better to iron out these details before proceeding too far. Once the article is written/revised to meet the periodical's requirements, it's a little late to start haggling over business matters.

Consider *payment*. As we'll see in Chapter 5, many technical periodicals do pay contributors, albeit modestly. In most cases, the editor will quote a price in his acceptance letter, usually for the entire article, but occasionally on a per printed page, per word, or (rarely) per column inch basis. Usually this is a firm rate. However, some editors will negotiate, especially if they'd like to commission a certain type of article. Since editors must work within often tight budgets, they are unlikely to offer you more, unless you ask.

Speaking of money, federal employees should take heed of certain restrictions in the 1989 Ethics Reform Act. Congress in its infinite wisdom has decreed that federal employees may NOT accept honoraria for any articles they should write, *even if the articles have no connection whatsoever with their work.* As of this writing (May 1992), several federal court challenges to this provision (Section 601c) of the Act have been lodged. Moreover, several bills have been introduced in Congress to delete the honoraria ban. By the time this book is published, either the courts or the Congress may have removed this silly prohibition. If they haven't, however, federal employees will have to (1) stop writing articles or (2) stop taking payment for them if they keep writing. In any event, civil servants and uniformed service personnel alike will have to tread carefully in this area. They will have to make sure that the article they've written or will write has been cleared by their agency's ethics officials. The alternative: risk a fine of up to $10,000. (Few articles pay that much.)

This brings up another matter, namely the distinction between "commissioned" and "speculative" articles. Commissioned articles are those that an editor assigns to an author whom he knows is an expert in a certain area and who can be counted on to deliver an article written to his specifications. If, like Harmon E. Interior (see Figure 1), you happen to be an ergonomics expert and a periodical editor happens to need an article on a certain subject that only you can deliver, then you may be asked by this editor to write it. In such cases, you have the advantage. The editor needs the article by a certain date and can get few others to write it. Hence, you have an excellent basis for negotiation. If you do not like the terms he offers, you can dicker — and you probably should. Although most negotiations center on money, other matters, such as deadlines and article content, also come up.

What if you should accept a commission, deliver the article, and learn that the editor will not publish it? It happens, believe me. About 5 years ago, I got an assignment to write a book review for a large, central North Carolina newspaper, one for whom I had written dozens of reviews. I wrote the review to the editor's specifications and turned it in. He sent it back asking me to make some changes. This I did, and returned it to him. He still didn't like it, but this time he simply said he'd changed his mind and wasn't going to run the review. Now, I had a decision to make: should I just forget about it and move on? After all, the review only paid $35. Or should I insist on my writer's rights and demand some payment for the article, even though it wasn't published? I did the latter. Bypassing the fussy editor, I wrote the publisher of the newspaper, explained the situation (emphasizing my long association with his paper), and asked for a $25 "kill fee" to compensate me for my troubles. I also enclosed a copy of the draft book review. He phoned me a few days later. "It's not our policy to pay kill fees," he began. "Maybe not," I replied, "but your editor assigned me this review, and I delivered. Besides, I rewrote it to incorporate his comments. I think I should be paid for my troubles." Maybe I convinced him, or maybe he decided that the amount wasn't worth arguing over, but at any rate, I received a $25 check from the newspaper a week later.

The moral here: be sure that you get the editor to agree, preferably in writing, to pay you a kill fee if he decides not to publish your article. (I must confess that I hadn't done this with the newspaper above. If I had, I would have avoided some extra work and unpleasantness.) If the editor doesn't agree to pay a kill fee, you should think twice about accepting an assignment from him; for you could very well be writing an article that never sees the light of day.

(However, I hasten to add that I've never had such an unpleasant experience with a technical editor. Those with whom I've worked have been unfailingly honest in their dealings. And although it is neither fair nor accurate to generalize about any profession, I venture to say that technical editors are, on the whole, easier to deal with than their trade publishing counterparts.)

If you write an article "on spec", you have no protection whatsoever. The editor of the publication has no obligation to you at all. He doesn't even have to read the article, though most editors will do so, out of courtesy alone. Then, why would anyone write an article on speculation? Perhaps you're an unpublished writer who wants to get your foot in the door at a prestigious periodical. Or perhaps you've already written most of the article in another form (e.g., a research report), and you wouldn't have to do much work to get it into publishable shape. Or perhaps, like my ill-fated book review, your article would not require much time to write. In any event, you would still be advised to query the periodical first. Even the shortest article takes more time and effort to prepare than a one-page letter.

WRITING YOUR ARTICLE

This will be a short section because I really have little to add to what so many others have already written. There are dozens of good "how-to" books on the subject of writing, technical and otherwise. These include: *Elements of Style* (Strunk and White), *On Writing Well* (Zinser), and *Effective Writing Strategies for Engineers and Scientists* (Woolston, Robinson, and Kutzbach), just to name three. Each discusses the nuts and bolts of the writing craft and presents techniques which, if used, will make your writing clear, concise, and comprehensive. I urge you to become familiar with them, especially the first. Or you may have a favorite writing text that you've used for many years. If so, get reacquainted with it. It doesn't matter which writing book you use. What's important is your finding one and using it.

Now it's time to start writing your article. More precisely, it's time to resume writing it, for you really started when you first began jotting down your thoughts about it. At this point, you may have to do additional research, to rework your outline, or perform sundry other tasks. But for simplicity, let's assume that you've finished all the preliminaries and are ready to begin writing.

Are there any "tricks of the trade" to make your job easy? Unfortunately, there aren't. Writing is hard work. It demands concentration, dedication, energy, and, above all, a positive attitude. Any book on writing will tell you that. (And the woods are full of them!) Although

I don't have much to say that's not already been said, I do have a few tips to convey — tips that have served me well and have made writing an enjoyable task.

Establish Good Work Habits

Most technical professionals wouldn't be where they are today if they hadn't cultivated solid work-study habits. They are accustomed to spending long hours at desks, in labs, or in clinics, concentrating on their tasks. They know what discipline means. If anything, many of them lean toward the opposite pole, workaholism.

Writing also requires a discipline which may or may not resemble your professional work habits. You may be a "desk jockey", but spend as much time away from your desk at meetings and seminars as you do sitting at it. And when you're in your swivel chair, admiring the view from the 9th floor window, how much time do you spend writing? If you're honest, you may find that your at-desk time is consumed by phone calls, reading, and other tasks which, though essential, aren't writing. To be effective, writing requires uninterrupted blocks of time. They needn't be big blocks — 15 to 30 minutes per session may be enough — but they should be continuous. Why? Because the Muse is a jealous, skittish creature that departs at the first sight of, or sound from, an interloper. And she is difficult to lure back once she flies away.

Therefore, *reserve a regular block of time for writing — daily, if possible, but at least weekly.* Evenings and weekends work for me, but so do lunch hours. I've been a desk diner for years. I just shut my office door, munch my brown bag lunch, read, and write. This not only gives me time to work, but also saves me money!

Set and Adhere to Writing Quotas

Every time you sit down to write, decide that you *will* write a certain number of pages before you quit. Most writers, successful and not so successful, do this. Stephen King writes six pages of double-spaced typescript every day, no matter what. Erica Jong's self-imposed quota is 10 pages of longhand. With yours truly, it's five or six longhand pages or three typescript, depending if I'm at the word cruncher. Now, I'll admit that that quota will seem easy to reach when the words are flowing. But on other days, it'll be a major struggle to produce one or two pages. My efficiency primarily depends upon two variables: my mood and my subject matter. If I'm "up", the pages fill up in no time; vice versa if I'm depressed. Similarly, if what I'm writing about is complex, I'm less efficient than if I'm conveying lightweight concepts. In any event, I try not to let these and other variables affect my session production.

Don't Stop Until You've Used up your Time or Met your Page Quota

If you write continuously, refusing to stop for anything, you will get more done in your session. This goes without saying. Less obvious is the fact that non-stop writing is often *better* writing. Non-stop writing is usually looser, less confined by the rules of grammar, spelling, and even logic. It allows you to tap your creative powers, to use the part of your mind that is suited to synthesizing theories, to finding connections among seemingly diverse concepts. The logical side of your brain will urge you to stop the writing to polish up your sentences, check your facts, to be logical and scientific. For now, ignore those sensible impulses and keep writing. Once you're finished, you can patch up what you've written (which is what I'm doing right now, by the way). In fact, the best time to do this is at the beginning of your next writing session. This allows you both to revise your most recent work and to warm up for the day's writing.

Make your Writing Good Writing

A U.S. Supreme Court Justice was once asked if he could define pornography. His response: "I know it when I see it." Good writing is like that. We know it's good when we read it. We may not be able to explain *why* it's good, we just *know* it is. Many writing teachers, not to say writers, have tried to identify the elements common to good writing. Some of these concern fiction — mood, tone, character development — elements that don't apply to technical writing.

But, others do. Put simply, good technical writing (1) has unity, (2) communicates well, and (3) is easy (if not enjoyable) to read. Let's take these elements one at a time. By "unity", I mean that every part of the writing is connected to the other parts, and that each of these parts contributes to meeting the objectives. Moreover, these objectives are reached in stepwise fashion. For instance, a book on the interstate highway system (see Chapter 4) wouldn't exhibit much unity if it contained chapters on the World Series, Paris fashions, or the Crimean War. Similarly, the book wouldn't be unified if a chapter on the history of road building was followed by one on road building methods, thence by another history chapter that takes up where the first one left off. That is, the parts must be relevant to the theme and presented in logical order.

The other two elements, *readability* and *communicability*, are related, yet have distinct identities. Readable writing contains several things that make it stand out. First, use of the *active* voice; it predominates. For instance, it's much easier to read "most historians agree that Edwin Drake drilled the first successful oil well" than "the first successful oil well was drilled by Edwin Drake, as agreed by most historians." The former sentence is vital and bright; the latter, reserved, anemic, and lustreless. Unfortunately, in most technical writing, the passive voice predominates. I can't say why exactly. Maybe technical writers feel that the passive voice is more formal, more professional sounding, and that it gives their writing a more authoritative ring.

Not true. Writing full of passive voice sentences is dull and imprecise. This doesn't mean that *every* sentence needs to be cast in the active voice. Sometimes the passive voice works better. (Example: "The innocent bystanders were hit by stray gunfire" reads better than "Stray gunfire hit the innocent bystanders.") However, in most cases, the active voice is preferable.

A second way to make your writing readable is to use shorter, simpler words, instead of the longer, ornate variety. You may think that fancy words sound nice and serve to enrich your writing. The fact is, they clog it up. They slow down the reader who is trying to derive some benefit from reading your writing. She is less concerned about *how* you say something than about what you're saying. Always remember: your readers have very little time to read, and the little time they do have must be spent well.

Third, along with preferring the plain word to the fancy, use more verbs and fewer modifiers, especially adjectives. Vibrant writing thrives on verbs. Verbs are to writing as a locomotive is to a train. A powerful locomotive will pull the train to its destination quickly and efficiently. Adverbs, which modify verbs, are analogous to the locomotive controls: they serve to regulate speed, modify direction, and add power when needed. Just as excessive controls on a locomotive can constrain its operation, too many adverbs can hinder the verbs from pulling their load. Use them sparingly.

Adjectives, however, are more like the cars on the train. To convey a load, a number of cars are required. Any more would be dead weight. A certain number of adjectives are needed to give the writing color and depth, but too many can slow the narrative to a crawl.

Fourth, vary the length of your sentences, while striving to keep them as short as possible. This serves to break the monotony of long, droning blocks of text which seem to go on forever... A few writers — notably William Faulkner — were well known for their interminable, convoluted sentences. Amazingly enough, their writing thrived as a result; but these geniuses are the exceptions. If you would imitate any writer's style, imitate Ernest Hemingway's or James M. Cain's. Their prose was spare, tight, unadorned. Granted, your technical writing won't be fiction (at least, we hope not!); but, you will have to keep your readers reading your writing. You don't want them to lay it aside just because it's too difficult to follow.

Fifth, sprinkle your text with figures, tables, and illustrations. These not only enhance the text, but also serve to break it up, to make it more attractive. Ours is, like it or not, a visual culture. The proliferation and immense popularity of images — photos, TV, movies, video games — has conditioned us to expect the same in our reading. As writers, we cannot change that (if indeed, we should want to). We can only acquiesce and include as many visuals in our writing as possible. For instance, instead of spending one or two paragraphs to explain the results of a series of measurements, compile them in a well-designed table or figure. Then, you can make the table/figure the foundation for the discussion and add what words you need to explain and interpret the data thereon.

Photos and other illustrations are also valuable. Sometimes they convey technical information, but often they do more: provide a bridge between the abstract world of the book and the real world it depicts. In a way, these pictures provide a "release" for the reader, in that the author doesn't have to waste mental energy trying to visualize an object or concept. Instead, he can spend his energy to interpret the technical discussion in the book.

Finally, take neither your writing nor yourself too seriously. Traditionally, technical writing is somber in tone. We come to expect that of it. Thus, when a technical book comes along that dares to be different, it is like a fresh breeze blowing through a smoke-filled room. A book like Octave Levenspiel's *Chemical Reactor Minibook* (Oregon State University Bookstores, 1979) illustrates this. Levenspiel, a renowned expert in the chemical engineering specialty of reaction kinetics, wanted not only to supplement his classic work, *Chemical Reaction Engineering*, but to have some fun as well. He was successful on both counts. His book is written with a light touch, making a rigorous subject easier and more enjoyable to learn. In my review of the book, I wrote: "[T]he author employs a rapid-fire style and a graphic format that is eye-catching, readable, and, above all, learnable. His equations and diagrams are peppered with notes and asides that are usually illustrative and often humorous. Humor is also evident in some of the wide array of problems he poses."[1]

The technical writing business needs more writers like Levenspiel, not so much to sell more books or periodicals, but to reach a wider audience. How do you give your writing the light touch? First, convince yourself that you're *enjoying* the work. If you can do this, the rest will fall in place. Before you know it, your writing will start to flow. You'll find yourself using clear, informal language — the active voice, shorter words, contractions. And you'll find yourself injecting witticisms, humorous asides, and, especially, anecdotes to enliven your prose. You may not be conscious of doing this, but you'll be doing it nonetheless.

Another mood-lightening device is the wise-saying-by-a-famous-person. I'm especially fond of them. Notice that most of the chapters in this book begin with a quotation from a well-known author. I've selected each quote carefully to best fit the chapter subject matter. These snippets provide food for thought, as well as a mini-introductions to the chapter. You'll find more quotes in Chapter 1, where the stage is set for the rest of the book, and in Chapter 7, where the final curtain is lowered. Incidentally, Chapter 7 also presents an in-depth background on, and advice from, prominent technical writers, editors, and publishers.

SANDING AND POLISHING (REWRITING)

Most writers yearn to reach the point where the first draft is completed. Having written several books and hundreds of articles during the past 15 years, I've experienced the intense relief and pleasure that comes with knowing that the first draft is finished at last. However, I've also experienced the let-down that comes when I think of how much more work is needed to shape the raw material into a finished product. True, most of the work has been done once the first draft is done. But recall the "80-20" rule: "Eighty percent of the work takes twenty percent of the time, and vice versa." (I think that rule was coined by some frustrated author, for it fits writing to a "T".)

The cleaning-up process begins, oddly enough, with your doing...nothing. That's right. Once you've finished your first draft, set it aside for a while. This gives it time to cool and you time to let your thoughts settle back into place. During this period, you may remember things that you've left out or should've included. Jot down these items now, while they're still fresh. Refer to this list whenever you return to the script. Otherwise, try to put the writing out of your mind. Involve yourself in other activities — easy enough to do, if you're as busy as most professionals. These activities might include other writing projects. Nothing wrong with that. The late Isaac Asimov, author of over 500 books, had several books going at any one time. When he was bored or stumped with one, he'd set it aside to work on another. But don't overdo it. Writing is demanding work, and if you engage in it too intensely during this cooling-off period, you may find yourself drained by the time you return to your manuscript.

When you do pick up the script, first read it thoroughly. Resist the temptation to begin editing at once. This read-through will give you an overview of the script to let you assess such matters as organization and content, and to identify gross errors such as repetitious material. Now is the time for renovation. If you've typed the manuscript on a word processor, you'll have a relatively easy time rearranging the text. However, if you've done it the hard way on a typewriter, get out your scissors and cellophane tape and start to cut and paste.

Once you've reorganized the script, begin editing in earnest. This involves taking each sentence, one by one, and doing the following:

- Rewriting ambiguous phrases
- Trimming excess words
- Correcting grammar and spelling
- Checking for consistency with tables, figures, etc.
- Verifying references
- Validating equations and calculations

This is exacting, time-consuming work, so allow plenty of time for editing. Don't assume that the editor will do it for you. Even if a copy editor is assigned to your script, there is no guarantee that she will do a thorough job. Moreover, she will not be as familiar with the material as you are, so the errors she would find likely would be superficial, not substantive.

THE FINAL COPY

By the time you've finished editing, your manuscript will probably have more lines and symbols than a 1940 Pennsylvania road map. The script would be in no shape to submit to the periodical. Clearly, you'll need a clean copy, either a retyping or a new printout. But before you do this, review the periodical's manuscript preparation requirements. Some of these are quite rigorous; others, not so. Regardless of their complexity, follow them as closely as you can.

Once you've retyped/printed out the article manuscript, proof it once again. If you need to make any more changes (and by this point they should be minor) mark them on the manuscript in dark ink. Use proofreader's marks to indicate the changes. (Most large dictionaries contain lists of these marks.) Your editor won't mind if your script is less than perfect, as long as it's legible and the number of changes is small (say, 1 or 2 per page). If, however, you need to make a lot of revisions, retype the changed portions.

Once you've gotten the article in final shape, photocopy it. Send a clean copy to the periodical, but keep the original. Unless the editor requests otherwise, send any drawings, photos, or other illustrations that comprise part of the article.

Finally, draft a brief transmittal letter to the periodical's editor and include it in the package. This letter should be brief, as what you have to say should be said in your article. Merely list what you are sending him, provide any explanations about the manuscript (e.g., special symbols used), and add anything else you feel is important.

WRAPPING THINGS UP

Once the editor receives your article, he will either begin editing it in-house or send it to peer reviewers. The latter will review the article and return their comments to the editor. As we discussed in Chapter 2, the peer review process can be grueling, both in terms of the time it consumes and the stress it can induce in the reviewer, the publisher, and, of course, the author. At the same time, the process usually benefits the article, as peer reviewers invariably provide valuable suggestions that would improve the piece.

Should you incorporate every one of these suggestions? Only if every one of them has merit. No editor worth his or her salt would expect you to blindly follow every peer review comment. However, you should be prepared to explain why you can't or won't incorporate a given comment. Your explanations, moreover, should be sound. (Just because incorporating a comment would require considerable time and effort on your part isn't justification for ignoring it.) In any event, if there is some question about a reviewer's comment, ask the editor to give you her name, address, and phone number, so that you can contact her. Once you've discussed the comment with the reviewer, you may find that the changes she is suggesting aren't as significant as you had thought. She may be willing to compromise a bit, especially once you've explained your side of the story. It is possible — and this happens surprisingly often — she may have misinterpreted what you were writing and didn't mean to comment as she did. No matter how your conversation turns out, you'll have found it to be worthwhile.

Once you've incorporated the peer reviewer comments, retype the article and send it to the periodical for the last time. Unless the editor has additional questions, you probably won't hear from him until the article appears in print, some months later. At this time, the editor probably will send you a "contributor's" copy of the periodical and several copies ("tear sheets") of the article itself. You have my permission to send the latter to your friends and colleagues. After all that you've been through, you're entitled!

REFERENCES

1.   Vatavuk, W. M. "Review of: *The Chemical Reactor Omnibook* and *The Chemical Reactor Minibook*," *Chem. Eng.* (February 11, 1980) pp. 11.

4 Getting Yourself Published: Part II — Books

Books as well as food nourish and warm people.
Books make connections.
May Sarton

In Chapter 3, we learned how to select, query, and get published in periodicals. Now, it's time to talk about books. Clearly, books and articles are horses of different hues and must be handled in different ways. Appearance is the most obvious difference. Most technical books are bound with hard covers, while articles are printed with others of like kind in softcover journals. This affords books a superior physical (if not pedagogical) advantage. In other words, they generally outlast articles, in more ways than one.

Before we go any further, let's review the periodical selection criteria presented in Chapter 3:

- Subject matter
- Type of article
- Targeted audience
- Circulation
- Readership
- Editorial policy

Would these criteria apply when selecting a book publisher? Certainly, "subject matter" would. Publishers do prefer certain kinds of manuscripts over others, though their wants are somewhat less selective than that of periodicals. So would "targeted audience", as most books are written with one or more segments of the technical community in mind. "Circulation", "type of article", and "readership" would not apply, as these measures of market penetration are not relevant to publishers. Instead, such statistics as total annual sales (in dollars, dollars per title, dollars per book category, etc.), as well as titles published per year and titles in print, would be more appropriate measures. Unfortunately, this kind of information is hard to obtain, as most publishers consider it proprietary.

Fortunately, some of this information is available for *a few* technical publishers. Perhaps the best-known source is *Writer's Market* (Cincinnati, OH; Writer's Digest Books). Updated annually, this moderately priced (under $25) reference lists market information for most of the "trade" (mass market) American publishers and a few of the larger technical publishers (e.g., McGraw-Hill). However, the data are usually limited to such statistics as the total number of books published annually and, in a few cases the number of submissions received annually. No sales figures are provided. Another reference (but a rather expensive one) is the *Literary Marketplace* (R.R. Bowker Co., New York). This massive tome lists the names, addresses, and senior personnel for every publisher in the country. It also contains listings for literary agents, editors, and other publishing professionals. Though the LMP contains little or no market-related data, it's valuable as a "publishing industry rolodex". (We'll cover market information sources more thoroughly in Chapter 5.)

A successful publisher will also get a lot of submissions, for writers will be drawn to it; however, success cannot be measured by the "bottom line" alone. A publisher's success is measured as much by its reputation in the technical community, as by its balance sheet. If a publisher is known to treat writers with respect during contract negotiations, to publish their books reasonably soon after they complete them, to market these books aggressively, and, most of all, to pay them honest royalties on time, then that publisher would be considered a success. Needless to say, if a publisher either cannot or will not do these things, his reputation among technical professionals will suffer.

BREAKING THE ICE (INITIAL CONTACTS TO BOOK PUBLISHERS)

If you have a book idea that you think would interest a publisher, by all means phone him to discuss it. Tell him why you think he should publish your book. Explain why what you have to say would be different from what others have written on that subject. Or if your book covers a subject never before explored, tell him why you think there is a market for such a book and outline that market. During this initial contact, also give the publisher some background about yourself — professional career, previous publications, special interests, etc.

As a result of this phone call, the publisher will tell you one of three things. First, he could say that his firm doesn't publish books on that subject. For instance, a firm that specializes in environmental titles wouldn't be likely to publish a financial analysis textbook. (However, there seems to be a trend toward diversification among technical publishers. Hence, a publisher is less likely to reject a solid proposal just because he's never published a book of that kind before.)

Second, the publisher could tell you that he's not interested in your book. This could be because he's already published, or contracted to publish, several other titles on that subject. Or it could be that he doesn't like your idea for some reason (logical or otherwise) and doesn't wish to pursue it further with you. If you get rejected either way, don't get discouraged. There are several other publishers you can contact. In fact, the publisher may suggest a few others for you to contact who may be interested in your idea.

Third, the publisher could say, "Yes, I'm interested. Send me a proposal." Now you've gotten your foot in the door. But it's just ajar. To open it all the way, you'll have to write a winning proposal. And that leads us to...

OUTLINES ET AL.

For most articles, a query letter will suffice for a proposal. A book proposal should provide more detail, however. Depending on the publisher's preferences, a book proposal might contain one or more of the following: topic outline, table of contents, preface, sample chapter. A book represents a major commitment on the publisher's part, as well as the author's. Among other tasks, a publisher must draw up a contract, set up a file, assign an editor to track the book's progress, and reserve space on his production schedule. Consequently, the publisher wants as much information as he can get before deciding whether to accept a book proposal.

From an author's standpoint, a book proposal might seem to be a risky undertaking, an investment of time and energy that may never pay off. It's not, though. In fact, the act of writing the proposal helps to focus your attention on the job at hand. Among other things, it helps you decide what you're going to write about and how you're going to organize your material. And you can use certain parts of the proposal (e.g., the preface) in the book. Here's a closer look at these proposal ingredients:

Outline

A book outline is a topic-by-topic listing of the contents, along with a few sentences to synopsize each. An outline for a hypothetical book on the U.S. interstate highway system might look like this:

THE U.S. INTERSTATE HIGHWAY SYSTEM:

A Socio-Technical History

- **The past as prologue:** This portion provides a capsule history of American roads, from the building of the Cumberland (National) Road in the early 1800s to the creation of the national highway system early in this century. It continues through the construction of the Pennsylvania Turnpike (the first superhighway) and of various state turnpikes and other toll roads built in the 1940s and 1950s.
- **Establishment of the interstate system:** In 1956, President Eisenhower signed legislation authorizing construction of a superhighway system to link together all of the lower 48 states. This system would be essential during national emergencies (e.g., to facilitate troop movements) and would also provide faster and safer travel for millions of Americans. These chapters focus on the political and social pressures that spawned the interstate system, such as the explosive growth in car and truck traffic during post-war years.
- **"Golden age" of the superhighway:** Most of the interstate highway system was completed during the 1960s and early 1970s. Although many interstates were built from scratch, many state turnpikes and other toll roads were simply absorbed into the system. (For instance, the Pennsylvania Turnpike became I-76.) This part of the book also talks about design innovations implemented by highway engineers to make them safer and more pleasant to drive (e.g., cloverleafs and scenic medians). Other desirable features of interstates are emphasized, such as time and fuel savings.
- **Energy crises and other roadblocks:** The "golden age" came to an abrupt end with the first Arab energy crisis in 1974. As these chapters show, energy and environmental pressures brought about changes to the way Americans drive, the most notable being the 55 mph speed limit. Though much maligned, the lowered speed limit proved to be as much a life saver as a fuel saver. Meanwhile, the trucking industry instituted such fuel-saving measures as twin semitrailers — vehicles that many automobile associations condemned as unsafe. In addition, environmentalists often thwarted construction efforts by requiring road builders to perform environmental impact studies before turning the first shovel. Finally, the escalating costs of fuel and highway construction made interstate highways less desirable, and many hundreds of miles went unbuilt. Still, during the 1980s, interstates were built, piece by piece, until by 1990, nearly 100% of the system had been finished.
- **I-2000: Interstates in the next millenium:** Interstates will have to adapt to American travel needs in coming decades. These chapters provide both a look at the current transportation scene and a peek at what will come. This includes status reports on the few highway projects yet to be completed, projections of future vehicle travel patterns and statistics, multiple uses of highways (e.g., mass transit lanes). We also focus on the escalating costs of maintaining highways, espe-

cially bridges. Finally, we present both sides of a controversial issue which the following question embraces: "Will interstates ever become obsolete?"

- **Appendices:** The following information is included in appendices: map(s) of the interstate system, road mileage figures by state, and construction and maintenance expenditures by year and state.

This sort of outline conveys, at a glance, the key information the book will contain. It provides just enough detail to let the prospective publisher know what you want to write about, but not so much that he feels overwhelmed. Notice that the topics the book will cover are presented in brief, informally written sentences, rather than in dry, pedantic lists. Note also that the outline is written in the *present* tense, not the future (e.g.,"This portion covers..."). This is not by accident. The present tense enlivens the outline, giving it an immediacy that few publishers would fail to notice. It gives them the message that this book is *already* written, at least in the author's mind.

Detail is also kept to a minimum. Excess detail not only makes the outline unreadable, but could also constrain you as you write your book. For while writing, your mind might open up avenues of thought that you were sure were "closed for repairs" when you prepared the outline. Put another way, outline preparation is an analytical, left-brain activity, while writing is, essentially, a creative process, controlled more by the opposite side of the cerebrum.

Table of Contents

Many publishers prefer to see a table of contents instead of a topic outline. Much like an outline, a table of contents lists the major subjects to be covered in the book, each of which corresponds to a chapter heading. The number of chapters in a technical book varies from 5 (brief monograph) to 20 or more (handbook). A chapter covers a major topic in depth and is subdivided, in turn, into several subtopics. The table of contents proposal should reflect this subdivision by listing the main chapter subtopics. This lets the publisher know, at least generally, what each chapter will cover, as chapter titles can be deceptive. Like the outline above, this helps you shape the book in your mind. In fact, an outline can be an excellent road map to follow when writing your way through the book. Figure 4.1 shows the table of contents proposal for this book.

Preface

In many cases, a publisher will ask an author to submit a preface for the proposed book along with a table of contents. The preface functions much like the first paragraph of the query letter described above. It tells the publisher (and eventually, the reader) what the book is about, who

54   *Marketing Yourself with Technical Writing*

GETTING PUBLISHED: A GUIDE FOR TECHNICAL PROFESSIONALS
Table of Contents (Revised 7/4/90)

Chapter 1.   Why *You* Should Write for Publication
- Advantages
  Professional development
  Personal satisfaction (e.g., discovering the value of your expertise to others)
  Financial rewards (albeit limited)
- Disadvantages
  Hard work (writing can be *very* difficult...)
  Time (...and quite time-consuming)
  Rejections (coping with same)
- What the rest of this book is about

Chapter 2.   Some Preliminaries
Before starting to write, a writer needs to...
- Discover his area(s) of expertise
- Determine if her writing would be an original contribution — or a rehash
- Investigate potential outlets for writing
  Periodicals (refereed and non-refereed)
  Books (texts, "how-to's", memoirs)
- Collect research materials
  Technical
  Non-technical (e.g., style guides)

Chapter 3.   Getting Published
- Finding potential publishers (books, periodicals)
- Writing queries
- Crafting an outline (book or article)
- Working with editors, publishers, and reviewers
- Special considerations for books
  Obtaining reprint permissions
  Reviewing proofs
  Assembling an index
  Publicizing the book (reviews, excerpts)

Chapter 4.   Technical Publishing: A Status Report
- Overview of the technical writing market
- List of magazines, journals, publishers (names, addresses, and terms)
- Other references (e.g., *Writer's Market*)

Chapter 5.   Legal and Business Considerations
- Copyrights
  Summary of 1976 law
  Exceptions to (e.g., government publications)
- Publishing contracts
  Written vs oral
  Common terms
- Taxes
  Income (deductions, etc.)
  Sales, business, other
- Libel and liability

Chapter 6.   Successful Technical Writers Have Their Say
Successful writers discuss their...
- Backgrounds — personal and professional
- Writing habits
- Writing advice
- "Philosophy of writing"

Figure 4.1. Sample table of contents.

it's written for, and how it will help him or her. A preface should be brief (two double-spaced pages, maximum), concise, and *alive*. By "alive", I mean that a preface should not only attract a reader's attention, but also stir his emotions, get him enthused, or at least interested in the book. In a sense, a preface is to a book what an advertisement is to a product. Without a hard-hitting ad to let customers know that a product is for sale, few would be sold.

A preface should start with a bang. One of the most hard-hitting technical book prefaces I've ever read began: "This book was written to help you make a lot of money as a freelance writer."[1] The book in question is Robert W. Bly's *Secrets of a Freelance Writer: How to Make $85,000 a Year* (New York: Dodd, Mead & Co., 1988).* Bly, a chemical engineer-turned advertising copywriter, is an extremely successful freelance writer who has written 15 books and hundreds of magazine articles. However, he has derived the bulk of his income from writing for businesses — mainly ads, sales brochures, and direct mail packages. Thus, he is a master at using words and images, first to arouse specific needs in individuals and then to convince them that the only way they can satisfy those needs is to buy his clients' products. (For more about Mr. Bly, see Chapter 7.)

The rest of the preface has to convey the meat-and-potatoes information just described, and it should twist the prospective reader's arm a bit. Bly's preface does this:

> This book is dedicated to the proposition that writers should be paid a fair dollar for a fair day's work...and that writing is a professional service worth the fees that other professions command... [I]f you feel frustrated by editors and publishers who seem to begrudge you every penny when it comes time to negotiate your fee or advance — this book can change your life!... After reading this book, you'll be in a position to earn big money from your writing, whenever you need it... You will have more money...more earning power...and the freedom, security happiness, peace of mind, pride, contentment, and self-esteem that go with it.[1]

Bly's preface closes on a rousing note, one calculated to encourage the purchase of his book. Your preface should do likewise. If the reader has read this far, she should know what your book's about, who it's for, and what she'll get out of it. Now all she needs is a sentence or two to nudge her into reading it. I wrote the final sentence of the preface to my book, *Estimating Costs of Air Pollution Control*, with this in mind, "Through all, I have tried not to lose sight of my goal: to help others in the air pollution control field develop sound, defensible (within ± 30%) cost estimates with a minimum of time and effort."

A word of caution, however. *Don't promise something in the preface that*

---

* Copyright © 1988 by Robert W. Bly. Reprinted by permission of Henry Holt and Company, Inc.

you don't deliver in the book. I can't imagine a writer being foolish enough to do that — on purpose. But something like this can happen, and very nearly happened to me. While writing the final preface for *Estimating...*, I came upon a sentence referring to material, an entire chapter in fact, that I had included in the original manuscript, but which the publisher and I agreed to delete before the final script was prepared. Needless to say, I struck out that renegade sentence at once!

For this reason alone, the preface in your query submittal rarely will be identical to the one that's typeset. You will probably rewrite it several times, to make it consistent with the book's evolving form and content.

DIGESTING FEEDBACK

Next, you would submit your book proposal to the publisher or, more than likely, an "acquistions editor" (A.E.), a staff person responsible for acquiring manuscripts for possible publication. The A.E. will, in turn, send your proposal to several reviewers. As described in Chapter 2, these "peer reviewers" will review the proposal and return their comments to the A.E. promptly, usually in less than a month. Some reviewers may prepare detailed critiques of the proposal, while others may simply jot their comments in the margins of the outline/preface/table of contents.

The A.E. will forward these comments to you at once. You are expected to consider them carefully and incorporate the valid comments into your proposal. (You'll be the one to decide which comments are "valid".) Once you've done this, submit the revised proposal to the A.E. If, in his opinion, the revised proposal is satisfactory, he will formally accept it, usually via letter. And this means that his firm will commit to publishing the book that your proposal describes.

CONTRACTS: TECHNICAL VS TRADE PUBLISHERS

As I said earlier, a book represents a major commitment both to the author and the publisher. Writing and publishing a book requires months of work and significant amounts of money. Consequently, it's mutually advantageous for the parties to formalize their business arrangement in some fashion. The means normally used is a contract.

The technical book publishing contract is much like its "trade" book counterpart, in that it specifies who will do what, when, and for how much, as well as such matters as who will own various publishing rights. Both types of contracts are also alike in that they're usually standard ("boiler plate") agreements drawn up by the publisher. However, the terms of these standard contracts are often negotiable, despite beliefs to the contrary.

Nonetheless, there are some noticeable differences between technical and trade book contracts. For one thing, technical book agreements are usually shorter and less complex. This is mainly due to the fact that "subsidiary rights" (movies, TV, etc.) rarely come into play with technical books; and because sales are usually much lower (alas!). Clauses such as the "reserve against returns", where the publisher withholds a portion of the royalties in case copies are returned, are rarely included.

Another difference concerns copyright ownership. As we'll see in Chapter 6, the 1976 Copyright Law decrees that the copyright for any created work is always owned by the creator of said work, unless he or she signs away that right. Thus, in nearly all trade book contracts, there is a clause that says, in effect, that the publisher will apply for a copyright for the book in the author's name. Not so with technical book contracts. Here, the copyright is almost always applied for in the publisher's name. Actually, it doesn't make much difference in whose name the copyright is registered. The only time it might is if a technical book were reprinted, say in a less expensive edition, or where a videotape was produced based on the contents. These kinds of subsidiary rights are rarely relevant, though others (e.g., foreign translation rights) are. A technical book is, 99 times out of 100, simply printed and sold. The copies remaining are either sold back to the author (at a discount) or just discarded/recycled.

COMING TO TERMS

"Copyright" and "subsidiary rights" are just two of the areas covered in a standard technical book contract. Here is an overview of a typical publishing agreement:

- **Introductory paragraph:** Lists the full names and addresses of the author and publisher and the date the contract was drawn up. (The effective date, however, is the date by when all parties will have signed the contract.)
- **Article 1 (publishing rights):** This article grants the publisher the "sole exclusive right" to publish all editions of the book throughout the world. It also lists the title of the book, to make sure that all parties understand what the book is going to be about.
- **Article 2 (manuscript preparation and delivery):** This covers specifications for preparing the manuscript (delivery date, format, number of copies, etc.), advances paid to author and when paid, and "publisher's escape clause", which says that he may terminate the contract — and demand repayment of advances — if the author fails to deliver an acceptable manuscript by the deadline.
- **Article 3 (permissions):** In this, the author agrees not to use any copyrighted material without first obtaining permission to reprint it.
- **Article 4 (copyright):** States that the copyright will be taken out in the publisher's name and that he will obtain it at his expense.

- **Article 5 (page proofs):** Simply says that the publisher will submit page proofs to the author who will correct and return them within a specified time.
- **Article 6 (royalties):** This article details the schedule of royalties and when they are to be paid. (Example: 10% on revenues from the first 750 copies sold, 12% on the next 750, and 15% on the rest; to be paid in 6-month intervals.) It also indicates that the same royalty schedule will apply to sales of future editions of the book as to the first edition. Finally, it states that the royalty rate for books sold at less than a certain price (e.g., 25% under the catalog [list] price) will be half of the regular rate.
- **Article 7 (review copies):** States that the publisher will send review copies to "important journals in the field" and that the author's recommendations will be followed when doing so.
- **Article 8 (author's copies):** In this, the publisher agrees to give the author a certain number of free copies of the book (10, typically) and to sell him additional copies at a discount (usually 30%) from the catalog price.
- **Article 9 (translation rights):** This says that the author and publisher will evenly split the revenues from the sales of rights to any language other than English.
- **Articles 10 to 13 (miscellaneous):** These brief articles indicate that the contract will be binding on the "heirs/successors and assigns" of the author (art. 10) and publisher (art. 11), that the contract will be governed by the laws of the state where the publisher is domiciled (art. 12), and that the royalties will be paid to the author (13).
- **Signatures and dates:** Of course, no contract is binding unless all parties sign it and date their signatures. If there are multiple authors, each must sign the agreement. Similarly, if more than one person is required to commit the publisher to the contract, each authorized person must also sign the contract. Finally, to satisfy our beloved Internal Revenue Service, the Social Security Number(s) of the author(s) must also appear here.

READING THE FINE PRINT

Let's focus on some elements of a technical book contract.

Advances

To an author, this is one of the most interesting articles of the contract. But it's not necessarily so to publishers. As a rule, technical publishers prefer not to pay advances. Their reasons vary. Some say that technical professionals are generally well-paid individuals who do not need advances for living expenses while they write their books. Although nominally true, this argument misses the point. Most authors (technical and otherwise) request advances because they would like some guaranteed return for their effort and expense. A book requires months, even years of hard work — research, outlining, writing, rewriting, and proofing — that could yield little or no return. Without an advance, an author is guaranteed nothing but a few free copies of his book and any enhancement to his professional reputation that it would bring him.

An advance lets the publisher share some of the risks involved with publishing the book. Granted, the publisher assumes risks of his own when he signs the contract. He must dedicate the time of his editorial and production staff to the project, pay for printing the book, and commit resources to marketing it. These goods and services do not come cheap, especially printing costs, which have escalated sharply in recent years. (To illustrate, a technical book costs from $5 to $10 per copy to edit, produce, and print, in printings of 1000 copies or more.) On the other hand, the publisher earns 5 to 10 times as much from the sale of a book than the author does.

Finally, an advance provides the author not only some compensation for his labors, but also insurance that his book will be published. One might argue that the author and publisher have signed a contract. Therefore, the publisher is commited to publishing the book. That's true — but only *if* the publisher accepts the manuscript. An author could produce a manuscript that is deficient, incomplete, or otherwise unacceptable to the publisher. According to the contract, he would have every right to reject it, and wouldn't even have to give a reason. The author would have little or no recourse if he did so. The publisher has invested little or nothing in the project. From his perspective, he would have little or nothing to lose if the deal fell through.

The author, however, would have invested his time and expense and would have next to nothing to show for it. On the other hand, if the publisher had paid the author an advance, he would have commited himself to the project. If the manuscript were unacceptable to him, he would be compelled to at least work with the author to improve it. To get a return on this advance, the publisher would want to market the resultant book more aggressively than if he had paid none. And that, in turn, would boost the book's sales.

How much are technical book advances, and how are they paid? As expected, the amount varies according to the book, the author, and the publisher. A first book by a first-time author might earn a nominal advance of $1000 to $2000 — enough to defray the writer's expenses but hardly to improve his lifestyle. At the other end of the spectrum, a prolific author whose books have sold consistently well might command an advance of $10,000, $15,000, or more. Though respectable, the latter amount is by no mean's huge, especially when compared to the multimillion dollar advances earned by bestselling novelists and nonfiction writers in the trade publishing world. In fact, smaller publishers cannot afford to pay five-figure advances, due to cash flow constraints. So, an author desiring a hefty advance might have to target his proposal to one of the larger publishers.

The payment schedule for an advance will also vary, depending on the terms of the contract. Typically, one half of the advance is paid when the contract is signed, the balance being paid when the author delivers an acceptable book manuscript. The operative word here is "acceptable". If the publisher does not accept the manuscript (and he has a lot of leeway here) then he does not have to pay the rest of the advance. In fact, he could have grounds to recover the advance money already paid. Of course, to do so he would have to sue the author. From the publisher's standpoint, litigation in such cases is usually impractical, due to the small amounts involved and the fact that the parties often reside in different states.

Suppose the publisher accepts the manuscript, publishes the book, but cannot sell enough copies to recoup his investment. Would the author have to return his advance? Rarely, if ever. By paying an advance, the publisher is investing in both the book and the author. The advance provides the author an incentive to write a *good* manuscript, not just an acceptable one, and to deliver it on time. He is betting that the book — the product — will be better and, hence, more salable. In most cases, the publisher will be right. If he's wrong, then he will simply write off the advance as a bad investment. And remember that his investment is not huge by any means, in most cases less than the cost of a midpriced business computer.

Royalty Statements

Keep in mind that the advance is an advance against sales *royalties*, not sales *revenues*. It must be "earned out" before the author will get any royalty payments.

My book *Estimating Costs…* is illustrative. For this book, I was paid an advance of $1000 when I submitted the complete manuscript. This amount was deducted from my first royalty check. (See Figure 4.2.) Again, royalties are calculated as a percentage of the sales revenue generated by the book. The sales revenue depends on two parameters: (1) book price and (2) number of copies sold. However, the revenue is not calculated by just multiplying the book's list price by the total number sold. As Figure 2 shows, my book was sold at average prices ranging from $50.18 to $59.60. The list price during the 4 months covered by this royalty statement was $65.00. If every copy had been sold at list, the royalty would have been $1846.00, not $1499.97.

Why are books sold at less than the list price? Simply, to boost sales, as lower prices stimulate demand. For instance, books are often discounted soon after publication, to give readers an initial look at the product. Some publishers participate in trade shows and exhibits where

Lewis Publishers

MR. WILLIAM M. VATAVUK   (WMV)
3512 ANGUS ROAD

DURHAM, NC 27705

Tax ID:

Royalty Statement for Period through 8/90   Page   1

Type	Level	Price	Sold In Period	Calculation Base	Royalty Rate	Share %	$ Earnings
Book: VATAVUK · EST. COSTS AIR POLLUTION (VTEC/142-4)							
(Lifetime net sales through 8/90: 284)							
-CHG		PREPAID ROYALTY	10/89				-1000.00
Rule: (4- 11)							
REG	1	50.18	197	9885.46 NET	10.00%	100.00%	988.55
REG	1	58.11	19	1104.09 NET	10.00%	100.00%	110.41
REG	1	58.65	45	2639.25 NET	10.00%	100.00%	263.93
REG	1	59.60	23	1370.80 NET	10.00%	100.00%	137.08

***** Book Total: 499.97

Figure 4.2. Sample royalty statement.

they display their recent titles. At one of these shows, a publisher may offer attendees a chance to purchase his books at a large (up to 25%) discount from the list price. This opportunity is typically limited to a 1- to 2-month period, after which the price reverts to the list. Figure 4.2 shows that 197 copies of the book were purchased during the first month at an average price of $50.18 each. Most of these sales were made in connection with a trade show held in late spring 1990.

However, notice how the average price steadily increased from months 2 to 4, while the monthly sales figures dropped. This reflects the increasing number of sales being made at or near the list (catalog) price. As this publisher derives much of his revenue from direct mail sales, the average sales price for a given month will be almost as high as the list. It would rarely be exactly as high, because a portion of the direct mail sales are made to bookstores at a modest discount (e.g., 10% off list). As Figure 4.2 indicates, the royalty for this period was based on the lowest percentage — 10% of revenues. Had more than 750 books been sold during this period, the royalty would have been calculated in two steps: 10% on revenues for the first 750 copies and 12% on the next batch, up to 1500. Royalties on the 1501st and succeeding copies would be based on 15% of the revenues generated by their sale. This is spelled out in the contract terms outlined above.

Do technical book sales ever reach that lofty 15% plateau? Some do; some don't. It depends on both the sales figures and the contract terms. As we'll see in Chapter 5, the "typical" technical book sells a total of 2150 copies over a 5-year period. However, according to the "typical" technical book contract, the maximum (15%) royalty rate doesn't kick in until the 3001th copy is sold. Conversely, many technical books fail to sell out the first printing (1800 copies, typically), so that they may never get beyond the 10% royalty level.

This may be discouraging, but compare these rates to those governing sales of trade nonfiction hardcover books. For these, the 10% rate applies to sales of the first 5000 copies. To generate royalties at the maximum (15%) level, a book would have to sell more than 10,000 copies.[2] But of course, the market for trade books is far vaster than the technical book market.

Delivery Dates

Someone once said, "They call delivery dates *dead*lines, because you kill yourself trying to meet them!" That doesn't mean that you must meet the contracted delivery date no matter what. However, you should take the delivery date seriously. You should plan your work not just to meet, but to beat the deadline, especially because you will have some say in setting it. During contract negotiations, the publisher will ask you how soon you can deliver the manuscript. At this time, you may be tempted to pull a date out of the air. Resist that temptation! Instead, ask for time to consider and then get back to the publisher with your proposed date. Meanwhile, force yourself to sit down and estimate the time it would take you to plan, research, and write your book. To this estimate, add additional time (e.g., 20%) as a contingency to cover unanticipated problems — illness, family and professional crises, and difficulties in obtaining data or secretarial services. Finally, when making your estimate, develop not just a single number, but a range of numbers, reflecting best- and worst-case scenarios. Such an estimate for the interstate highway book described above might look like this:

Activity	Calendar time to complete (weeks)
Work plan	1–2
Library research	3–5
Contacts with USDOT, etc.	4–6
Writing (6 chap. @ 4–6 wks.)	24–36
Rewriting	2–4
Prepare final manuscript	3–5
	Total: 37–58

Adding 20% to this total yields a total of 44 to 70 weeks or about 10 to 16 months.

Keep in mind that this is a calendar time estimate. It should reflect the fact that writing this book will not be the only thing you'll have to do in the next year or so. Most technical book authors write only in their spare time. In fact, you may spend only 10 to 20% of your waking hours on this project.

When you get back with the publisher, tell him that you'll need $1^1/_2$

years to write the book, and suggest a delivery date consistent with that. Thus, you'll have given yourself even more time to deliver the manuscript than your worst-case estimate. Don't feel lazy. Most of the time, when we estimate the time needed to do something, we tend to underestimate it, not the opposite. So, your "pessimistic" deadline may turn out to be rather optimistic.

The publisher will probably agree to your proposed delivery date, unless it's unreasonably long or the book contains dated information (e.g., cost data). If he doesn't agree, you can either trim your estimate or trim some facet of the project (e.g., research) to meet his deadline. Remember: your primary objective is to get your book published, not to squeeze an extra month or two from the publisher.

But what if something comes up and you find that you absolutely cannot meet the deliver date? Should such a case arise, contact the publisher at once. Explain your situation, providing a *good* reason for the delay. Most publishers are reasonable people who are used to the "give and take" of publishing arrangements. They will probably give you an extension if it's justifiable and not too long. However, don't count on more than a 1-month reprieve, as the schedules for many related activities — production, printing, marketing, et al. — depend on when you deliver the final script.

OUTSIDE ADVICE

Even after you negotiate with the publisher, some parts of the contract may still be unclear or unacceptable to you. If so, you may want to ask an attorney to review it. He will examine the contract terms and meet with you to advise you about them. If necessary, he can contact the publisher to clear up any remaining ambiguities or misunderstandings. He may even be able to get you better terms. Even if he can't, you will at least feel less anxious about signing the contract once your attorney has scrutinized it. But wouldn't the publisher resent your bringing a third party into the negotiations? No. In fact, he may come to respect you more for having done so. In his eyes, you will seem more professional, a person who takes his writing seriously. Should you always consult a lawyer? No, again. Although some writing guides say you should, I've found that an attorney is worth retaining only if there are contract problems to resolve. If the contract negotiations have been concluded to everyone's satisfaction, you don't need one.

FINAL DETAILS

All right. With or without a lawyer, you and the publisher have come to terms and are ready to formalize your working arrangement. Before signing it, read your contract one last time, to be sure that you understand every part of it, and that it says what you expect it to. The publisher will have sent you two copies of the contract, both of which he has already signed. If the publisher is a small firm, just one signature will appear on it. However, if it's a larger company, especially a subsidiary of another firm, two or more signatures may be needed to make it binding upon the publisher. In any event, only your signature is needed to bind you to the agreement. No notarization or witnesses are required. Don't forget to date the contract, also. The day you sign becomes the effective date of the contract. For many reasons, this date could very well prove to be crucial, especially if (heaven forbid!) you should ever find yourself in litigation.

There is another bit of paperwork you should complete before you begin working on the book in earnest; and that is the "author questionnaire" (A.Q.). Developed by the publisher's publicity department, the A.Q. requests information about the author's professional and personal background that would have some bearing on the book. Other questions are more probing, for example, "In what ways does this book make a unique contribution to the field of _____?" Clearly, they require more than one- or two-word replies. Spend as much time as you can answering these questions. The more information you give the publicists, the better they can promote your book. The A.Q. may also request a photograph, usually a black-and-white, wallet-size portrait pose. (Color photos are more difficult to reproduce.) A snapshot is acceptable, but you'd be better off having a professional photographer shoot you. After all, this is the photograph that will appear in the book, and you want to look your best. In addition, you can order extra prints of the photograph that you can use for other purposes.

Some writers prefer to complete the A.Q. after they've finished the book. But I've found it best to complete it before writing the book. It helps give me the "jump start" I need to get going.

WRITING YOUR BOOK: SOME WORDS OF WISDOM

In many ways, writing a book is like writing an article or other short piece. The same rules of grammar apply, as do the same writing tips. (I discussed the latter in Chapter 3. I invite you to get reacquainted with them now.) However, there are some differences worth mentioning. Please keep them in mind when writing your book.

The first is the most obvious — *length*. Since a book is much longer than an article, it takes much more time to write. That translates to much more time spent at the typewriter/word processor. It also means more time for research because that additional verbiage will require a sturdy factual foundation.

Second, a book requires more *organizing* than does an article. When writing an article, you can nearly keep an outline of it in your head. At most, you'd need a page to list the main and subtopics. A book requires much more. You'd have to prepare a detailed outline, refer to it regularly when writing, and be ready to revise it when necessary. Writing a book without following an outline is like trying to drive through West Virginia without a road map. It's a sure-fire way to get lost.

Along with an outline, you'd need to keep track of the tables, figures, and references by chapter, so as to avoid duplications, omissions, and other confusion. For instance, while nearly finished writing a chapter, you might need to reference data obtained from a published source. Immediately, you ask yourself, "Have I already cited this reference earlier in the chapter? And what was the number I gave to the last reference cited? Was it 17 or 18?" Granted, these kinds of pesky questions can crop up when writing an article as well, but it is much easier (and less time consuming) to page through a 10- or 20-page article than to dig through a 50- or 60-page chapter. The "search" and other features of some word processing programs make these chores easier to do, but they don't relieve us from doing them altogether.

The third, and perhaps least appreciated, difference between writing an article and book is what we might call the *"commitment factor"*. There is a certain amount of inertia associated with beginning any writing project, even one as small as jotting a thank-you note. But once you overcome that mental inertia and "get the flow", you proceed steadily, knowing that the amount of writing to be done won't take forever to complete. In fact, you may even be able to complete the project in a single sitting. Your initial burst of energy may be strong enough not only to overcome the inertia, but to carry you through to the very end. Thus it often is when writing an article. The "light at the end of the tunnel" is very easy to see and glows ever brighter moment by moment.

A book is another matter altogether. Before you start a book, you know that it's going to require a major commitment from you — months, perhaps years of hard work. Your initial enthusiasm might carry you through the first one or two chapters, but after that it usually peters out. And then the going gets tough. If you expect to finish the book, you'll have to grit your teeth and grind out the pages, one by one. Although this is hard, demanding work, there are a few ways to make it a little more bearable: always keep your "eye on the prize", break the big job into small, manageable tasks, and avoid being a perfectionist.

Eye on the Prize

When things seem especially dark, and you don't think that you can bear to write another word, just close your eyes for a moment and visualize yourself holding the completed book in your hands. Slide your hands across the cover, riffle your fingers through the pages, smell the bracing odor of a newly printed volume. Then, with pride, gaze upon your name on the cover — the cover of the book that *you* have written. Now open your eyes, and say, aloud, "Soon that daydream will be a reality." Finally, get back to your writing machine!

Break up the Big Job

That is, instead of thinking, "I have a book to write," say to yourself, "I have this section of Chapter 4 to finish." And then concentrate on completing that section. Pretend that it's the most important thing you have to do and focus your energy on it and it alone. Don't even think about how you'll write the rest of the chapter, let alone the rest of the book. This form of self-hypnosis (for that's really what it is) will not only make the writing easier, but more enjoyable, as well.

Don't Be a Perfectionist

Writers have a saying: "A perfectionist can write an article, but only a *pragmatist* can write a book." Anyone who has written both will understand what this means. When writing a short piece, you have ample opportunity (and temptations) to indulge in perfectionism. You can draft a sentence, redraft it, and keep doing so until you're satisfied with its form and content. Similarly, you can check and recheck numbers in tables, scrutinize figures to see if they are accurate and aesthetically pleasing, and verify citations against original sources. You can indulge in perfectionism simply because your work is short and, compared to a book, cut and dried. For example, you don't have to bother with such things as making an index or obtaining reprint permissions.

But when writing a book, perfectionism is a dispensable luxury. You have neither the time nor the energy to fuss over a 50,000-word manuscript as you would a 3000-word article. What's more important, if you tinker with the prose too much, you can get yourself so bogged down in details that you can no longer tell whether you're getting your message across. That is, fussed-over sentences may sound better when they stand alone, but how well do they fit into the paragraph and how well does this paragraph connect with the ones that precede and follow it? Keep in mind that someone reading a book isn't going to be inclined to dwell on every single sentence. Rather, she wants to get the gist of what's being said; she wants to grasp the "big picture". If she wanted to savor beautiful words and phrases, she would read poetry.

Not being a perfectionist does not allow you to be sloppy, however. You are still obligated to check your facts and to present them so clearly that there is little or no chance they will be misinterpreted. But despite all your hard work, errors will creep into your book like proverbial thieves in the night. Just accept that humbling fact. Either you or your readers will find them, sooner or later. The best you can do is compile these errors and correct them in the next edition.

Keep these three things in mind and follow the writing tips discussed in Chapter 3. They will serve you just as well when you write a book as when you write an article — or anything else, for that matter.

THE FINAL MANUSCRIPT

There is another big difference between books and articles: the ingredients. Unlike an article manuscript, which typically includes only the text, figures, tables, and references, a book manuscript has more components.

Title Page

This is an $8^{1}/_2 \times 11$ sheet of paper that lists the book title, subtitle (if any), and your byline, name (if different from your byline), address, and phone/fax number. Figure 4.3 is an example. Notice how the information is centered on the page.

Table of Contents

This should list both the chapter titles and the major topics covered in each. Recall that you included a table of contents when you developed your book proposal. Take a moment to compare that table to the contents of the manuscript. Chances are, there are more than a few differences between the two. This shouldn't be surprising, for your book has evolved while you were writing it. Although the gist of the material covered hasn't changed appreciably, the *shape* of it undoubtedly has. For instance, I had originally intended to cover book contracts in Chapter 6 of this book, but while writing, I found that contracts fit better in this chapter. For illustrative purposes, here is a sample table of contents for Chapter 2 of my book *Estimating Costs of Air Pollution Control*.

MARKETING YOURSELF WITH TECHNICAL WRITING:

A GUIDE FOR TODAY'S PROFESSIONALS

William M. Vatavuk

3512 Angus Road

Durham, North Carolina

(919)-541-5309

Figure 4.3. Sample title page.

Unlike this example, your table of contents will not have any page numbers. The page numbers won't be determined until after the "page proofs" have been made. (We'll discuss page proofs later in this chapter.)

Preface
As described above, these two or three pages summarize what's in the book, who it's written for, how and when it should be used, and so

forth. This will be the final preface, so make sure you don't omit anything or, worse yet, include something that shouldn't be there.

Acknowledgments, Dedication, et al.

These items are optional; include them if you like. I think that an "acknowledgments" page is a nice touch. For one thing, it gives proper credit to those "without whose help the book could not have been written". Moreover, it displays the breadth and depth of your research and contacts.

In the acknowledgments, you may want to list individuals, firms, or both. If research facilities, such as libraries, have been helpful, include them too. List them alphabetically, in order of importance, or in any way you prefer.

The dedication — "the page that no one ever reads" — also adds decoration to the package. It's traditional for authors to dedicate their books to their long-suffering spouses (or, these days, their "significant others"). Many do so sincerely, though others no doubt do it out of loyalty, for appearance' sake, or just to keep peace in the family. Other dedications are used — to friends, mentors, groups, institutions, humanity in general — all of which are appropriate. This is, after all, your book.

WRAPPING THINGS UP

Once you've finished writing and have printed out all parts of your manuscript, proof it once again. If you need to make any more changes (and by this point they should be minor) mark them on the manuscript in dark ink. Use proofreader's marks to indicate the changes. (Most large dictionaries contain lists of these marks.) Your publisher won't mind if your script is less than perfect, as long as it's legible and the number of changes is small (say, one or two per page). If however, you need to make a lot of revisions, retype the changed portions.

Once you've gotten the script in final shape, photocopy it. The number of copies to make will depend on your contract; however, two is typical. Send these copies to the publisher, but keep the original. *Never* send the original, especially if the script contains drawings, photos, or other illustrations that you could not reproduce with your typewriter or word processor. Eventually, the publisher may need to have the original illustrations, but this won't be until the book is in production. Before that, the book will have been peer-reviewed and probably revised, at least to some extent.

Some publishers also ask you to send a copy of the book on computer diskette(s). There are two reasons for this. First, editors find it much easier to edit a manuscript electronically than manually. Also, recent advances in publishing allow a typesetter to produce "page proofs" directly from diskettes. If you have to send diskette(s) of your script, first find out if the publisher's word processing program is compatible with yours. If it's not (the usual situation), convert the diskettes to "ASCII" format, so that they can be read by the publisher's equipment.

Finally, draft a brief transmittal letter to the publisher and include it in the package. This letter should be brief, as what you have to say should be said in your script. Merely list what you are sending him, provide any explanations about the manuscript (e.g., special symbols used), and add anything else you feel is important.

Mail the diskettes *separately* from the manuscript, using special diskette mailers. This is a sensible precaution to take. (Anyone who has done business regularly with the postal service can relate his own horror stories of lost mail.) In addition, spend a few dollars more and send the diskettes and manuscript "certified with return receipt". This service costs $2 per item, in addition to first class postage. Should the item become misplaced, the ever-industrious postal service could trace it through the certified number (or at least try to). For more security, you can insure the items. (See your Postmaster for more details.)

THERE'LL BE SOME CHANGES MADE

The next few weeks and months might seem to last forever. You've finished your manuscript and, in your view, there's nothing more to do until the box of contributor copies arrives. The fact is, there is a lot going on — at the publisher, at the printer, and, most importantly, at the peer reviewers. Your manuscript may be finished, but it's not *final*; and it won't be final until the reviewers' comments have been incorporated or otherwise dealt with.

As I discussed earlier, the peer review process can be grueling, both in terms of the time it consumes and the stress it can induce in the reviewer, the publisher, and, of course, the author. Many writers consider their work to be their best. Ergo, it cannot be improved upon. Reviewers' comments, however well conceived, cannot be incorporated into the manuscript, for it is already "perfect", after all. If you harbor such misconceptions, rid yourself of them at once! If you don't, you may wind up with a finished manuscript that never becomes a book. Recall that clause in your contract: "Upon acceptance of the final manuscript, the Publisher will..." If the publisher doesn't accept your manuscript, he won't publish your book. It's as simple as that.

Does this mean that the publisher will reject a manuscript for any reason, however trivial? Hardly. He can't afford to. A publisher who rejects every less-than-perfect manuscript he contracts for won't publish very many books, and he won't stay in business for long. In other words, publishers and authors need each other. However, a publisher could very well reject a manuscript that fails to meet his minimum standards, especially if the author refuses to cooperate in improving it. This is where the peer reviewers come in. They provide the "catalyst" for this process. Without their expert input, the publisher would be severely hampered in his efforts to improve manuscripts.

This is how the process works. Soon after you submit your finished manuscript, the publisher mails it to two or more peer reviewers whose expertise matches the subject matter of the manuscript. Each reviewer is allowed 2 to 4 months to review the manuscript (or perhaps just a portion of the script) and return her comments to the publisher. The comments may be in the form of a letter or just marked in the margins of the script. Once the publisher gets each set of comments, he forwards them to you for your consideration. Notice that I didn't write "for your incorporation"; for at this point, the publisher wants you to respond to the comments — to tell him which ones have merit and which don't, and why. If you agree with all of the comments, by all means incorporate them into a final script. You both will benefit from disposing of the changes promptly.

What if the comments are not to your liking or they request so many large-scale changes that it will require months of your time to incorporate them? In the first case, you have to decide whether you can live with the changes. You have to ask yourself, "Am I refusing to make these changes because they're invalid or just because I'm too proud to make them?" If incorporating the changes would damage the script, explain this to the publisher and provide evidence for your position. Offer to arrange a meeting or conference call with the reviewer so that you can iron out your differences. You may find out that the reviewer merely misinterpreted what you wrote. If you and the reviewer can't agree on the comments, ask the publisher to send the script to another reviewer. If the publisher is convinced of your desire to optimize the quality of your script, he should agree to do this.

A different situation arises when the comments are valid, but require so much additional work to incorporate that it would delay the publication schedule by months, even years. In this case, both you and the publisher may conclude that the comments just aren't worth the time and effort to incorporate. This is especially the case when the book's subject matter has a short shelf life. Delaying publication to incorporate extensive, though valid, comments could make the book out-of-date before it's printed.

Suppose, however, that there really is no hurry to get the book into print. Should you take the time to incorporate extensive comments then? Again, that's up to you and the publisher. In the long run, it might be better if you grit your teeth and make the revisions. The book would be better and would probably sell better.* Even if it didn't sell better (and you'd have no way of knowing for sure anyway), at least you'd have had the satisfaction of knowing that your book reflects the comments of some of the leading professionals in the field. There's something to be said for that.

LAST-MINUTE TASKS

At last, the changes have been incorporated into the script and your book is ready to go into production. Now, you can finally settle back to relax and await the arrival of that fabled box of contributor's copies. Don't stretch out just yet. There are still reprint permissions to obtain and editor's questions to answer.

Reprint Permissions

In most cases, material in your book that you've taken from other published sources can be referenced in the normal way, via footnotes, bibliographies, etc. However, if you've appropriated a large block of information — a table, figure, or more than a small amount of text — you'll need to obtain a reprint permission from the holder of the copyright on the material. This could be a publisher, a periodical, a company, or any other entity. You may also need to get permission to reprint a photo, drawing, or other illustration. In fact, as we'll see in Chapter 6, the material doesn't even have to be published to be copyrighted.

The most efficient way to do this is to send the copyright holder a permission form to sign. (Actually, send him two copies, one for him and one to return to you.) Provided by the publisher, this form describes the material to be reprinted, and where, how, and by whom it will be used. (Figure 4.4 depicts a sample form.) Be sure to obtain these permissions well before the book is published. Send them to the publisher as soon as you collect them all.

Editor's Queries

As soon as your manuscript is "accepted" by the publisher, he will assign a review editor (in-house or contractor) to work with you while the book is in production. She will go through the manuscript with a proverbial fine-tooth comb, to spot errors, inconsistencies, and other vexations. These she will correct. She will also edit the script for con-

---

* I say "probably", because there is no hard-and-fast correlation between quality and sales. Some inferior books have sold very well, while some excellent books have gathered dust in publishers' warehouses. That's what makes publishing at once exasperating and fascinating.

LEWIS PUBLISHERS, INC.   PHONE (313) 475-8619

121 SOUTH MAIN STREET ☐ P.O. DRAWER 519 ☐ CHELSEA, MICHIGAN 48118 ☐ FAX: (313) 475-8650

PERMISSION REQUEST FORM

I am writing to request permission to reprint the following material from your publication:

Author/Editor _____

Title _____

Publication Date _____

Page No. _____   Figure No. _____   Table No. _____

This material is to appear in the following work which Lewis Publishers, Inc. is currently preparing for publication:

Title/Author/Editor: William M. Vatavuk, Author, *Marketing Yourself with Technical Writing*

This book is scheduled to be published in 1992.

I am requesting nonexclusive world rights to use this material as part of my work in all languages and for all editions. If you are the copyright holder, may I have your permission to reprint the material described above in my book. Unless you request otherwise, I will use the conventional scholarly form of acknowledgement, including author and title, publisher's name, and date.

If you are not the copyright holder, or if for world rights I need additional permission from another source, will you kindly so indicate.

Thank you for your consideration of this request. If permission is granted, please send this signed form to me at the address listed below. If permission is denied, please inform me in writing.

Yours sincerely

_____   _____
(contributing chapter author's name)   (signature)

_____
street address   city   state   zip

****
****

The above request is approved on the understanding that full credit will be given to the source.

Approved by: _____   Date: _____

Company: _____

Figure 4.4. Sample permissions form.

sistency with the publisher's style guidelines. During this process, she will undoubtedly have some questions for you. Some of them might seem trivial to you, but to her (and the script) they're important. By all means, cooperate with her. Make, not just take, the time to answer her questions, even if that means digging through your references to verify a technical point or datum. Good editors are worth their weight in platinum and should be treasured accordingly.

In addition to editing the script, she will also work with the art department in designing the cover for the book. Ask to have input in this process. She may also ask your opinion on such matters as paper stock, typeface, and binding type (cloth or "casebound", a plastic laminate). If you do, take advantage of the opportunity. For, despite what you might think, the appearance of a book is nearly as important as its contents. A nicely designed, well-constructed book is pleasing to the eye and hand, more durable, and, consequently, a pleasure to use.

THE PROOF IS IN THE PAGES

Once the editor is finished editing the manuscript, she will send it to a typesetter. The typesetter develops "page proofs" from the manuscript. As the name implies, page proofs are mock-ups of the book pages and tables, but without the figures, photos, and other artwork.* The size, typefaces, layout, and numbering of these pages are identical to what will appear in the published book.

Your contract probably obligates you to review these proofs, mark your changes on them, and return them to the publisher within a specified time (typically, 30 days). Even if you weren't required to review the proofs, you'd be foolish not to. Despite your and the editor's best efforts, errors will have crept into the proofs, many of which only you, the author, can identify. These include "technical" errors in equations, tables, figures, and the like. Typically, the editor will send you two copies of the proofs. One copy is for you to mark up and return, while the other is for your files. In addition, you'll be using the second copy to make the index. The editor will retain a third copy which she and or a copy editor will review simultaneously. In this way, the proofs will get two reviews, a practice that allows fewer mistakes to slip through the cracks. (As I said above, some errors will survive this gauntlet, despite your and your editor's best efforts. However, a lot more would make it, were it not for the page-proofs review process.)

However, keep in mind that the page-proofs step is the last one before publication. Now is *not* the time to make major changes to the

---

* Before the advent of computerized typesetting, the publisher would generate "galley proofs" of the manuscript. These were $8^1/_2 \times 14$-inch sheets that the author was obligated to review and correct. From these corrected galleys, the typesetter would generate page proofs. Though reviewing galleys was extra work for the author and editor, this step allowed them to catch errors they otherwise might have missed had they reviewed page proofs alone. But the computer has allowed publishers to do away with galleys, thus saving them time and money. However, one can argue whether deleting this step has resulted in better books being published.

book, especially significant additions, deletions, or rearrangements. Large-scale revisions at this stage would cost the publisher a tidy sum — one which he could justifiably expect you to pay. From the publisher's viewpoint, you've had plenty of time before this point to revise the script; if you still need to make any major rewrites, he shouldn't pay for it. For this reason alone, your revisions to the proofs should be editorial. Any material changes should be no longer than one or two sentences each. Of course, your publisher may have a stricter (or more lenient) policy regarding the extent of changes permissible. A copy of these guidelines should be included in the package containing the proofs. Also in this package should be instructions for correcting the proofs (e.g., proofreaders' marks to use).

When you've finished reviewing the proofs, return them to the publisher, along with original illustrations, reprint permission forms, your author photo (if one is used), the author questionnaire (if you haven't already submitted it), and any other items your publisher needs to collect before publication. You may wish to send this package insured, or at least via certified mail, to assure its safe arrival. Some of the contents, especially the illustrations, may be very difficult to replace.

INTO THE INDEX

A technical publisher I know has displayed his books at trade fairs for more than 20 years. At a typical fair, browsers stop by his booth, pick up his books, and thumb through them. "They look at two things in a book," he said, "the table of contents and the index. In my experience, those are the things that sell a book."

What this publisher said makes a lot of sense. Is there a quicker or easier way to find out what a book does or doesn't contain? The table of contents tells you what topics the book covers and in what order. The index, however, gives a "between-the-lines" look at, not just the major topics, but the *variety* of subjects covered in the book. Moreover, the index indicates the *depth* to which each subject is covered. A table of contents might tell us that a topic — superhighway construction materials, for instance — is covered in Chapter 6 on pages 98 to 102, but the index may show that related topics are treated on pages 75 to 79, 151 to 157, and 204 to 205 also. Hence, the index is important to the book and should be developed with care and thought. It's not just a useless appendage tacked on as an afterthought because every technical book wouldn't seem complete without one.

The two types of indexes (or "indices") most commonly used are those arranged by "name" and by "subject". As the word indicates, a "name" index is an alphabetical list of names keyed to the page(s) on which these individuals are named and discussed (e.g., "Herman,

Woodrow Charles, 47 to 61"). Though useful for many nonfiction books (e.g., biographies), a name index is of limited utility for a technical book, because the subject matter is oriented more toward thoughts and things than toward people.

Conversely, a "subject" index is an alphabetical list of key words or word groupings, each representing a topic in the book. A subject index will likely include a mixture of listings: "one-tier" (topics only), "two-tier" (topics and subtopics), "three-tier" (topics + subtopics + sub-subtopics), and so on. The detail of the index depends on the material being indexed and on the amount of time and effort the indexer is willing to spend. Here's a sample of a two-tier index entry:

> Mist eliminators. *See also* Wet scrubbers
>   equipment costs, 137
>   sizing parameters, 124, 126–128, 130
>   types, 124

In this example, there is also a cross reference to a related topic, "Wet scrubbers." This serves to direct the reader to the second entry, where he will see...

> Wet scrubbers
>   annual costs, 138–139
>   auxiliary equipment (pumps), 138–139
>   capture mechanisms, 120–121
>   equipment costs, 134–138
>   sizing, 126–134
>   total capital investment, 138
>   types, 120–126

Clearly, this second entry covers more ground than the first, which deals solely with a single type of wet scrubbers, mist eliminators. However, there is some overlap, especially in the "equipment costs" and "sizing" subtopics. A thorough index makes liberal use of cross-referencing, a technique that serves to unify the index and, in turn, the book.

How do you develop an index? One way is to hire a professional indexer. This frees you from several hours of tedious work. However, their services aren't cheap: $1.50 to $6.00 per printed book page.[3] Or, you can do it yourself. Though making an index is exacting work, it's not an overwhelming task. A technique that has worked well for me involves the following steps:

1. Using a clean copy of the proofs, go through each page carefully and mark the key words (or phrases) with a highlighter.

2. With a package of lined index cards at hand, go back through the proofs, mark a key word on a card the first time you see it. Also jot down the page number(s) on which the key word appears.
3. Depending on the first letter of the key word, place the new card in one of six stacks, as follows: "A–E", "F–J", "K–N", "O–R", "S–T", or "U–Z".
4. As you come upon a key word, check the appropriate card pile to see if you've already made a card for it. If you haven't, make one. If you have, add the corresponding page number(s) to the card. Sometimes, you may have to add one or more subtopics to a key word card, such as "equipment costs" under "Wet scrubbers" in the illustration above. As in this illustration, you may also find that some of the page numbers will overlap. This occurs when the discussion about more than one key word topic or subtopic is found on the same page(s). But all this is not as complex as it sounds. Once you "get into" the indexing, you will find things falling into place in short order.
5. Once you've written your last entry, arrange the index cards in alphabetical order, one stack at a time.
6. Type the index (double-spaced).
7. Send the index (on both hard copy and computer diskette) to the publisher. Again, you're advised to send the material certified, insured, or some other way to protect it.

Your publisher may ask you to follow his own directions for preparing the index. If these should differ from the six steps above, by all means ignore them and follow your publisher's guidelines. However, if your publisher has no preferences, these steps should prove to be helpful.

POST PUBLICATION

In the trade publishing world, the publication of a major book is a big event. Parties, press conferences, book signings — all well publicized, of course — are held to promote its debut. Unfortunately, these things rarely happen when a technical book is published, even if it's a "star" title. For one thing, technical book publishers aren't equipped to promote any one of their titles to such an extent. Their promotion efforts are usually limited to placing advertisements in a few technical journals, sending review copies to the relevant periodicals, circulating flyers announcing the book, and listing the book in their catalogs. In addition, as I mentioned earlier, publishers do promote their titles at trade fairs and technical meetings. At these fairs, they might have some of their better-known authors present to sign books. Despite their trouble and expense, fairs and meetings do help to sell books.

Another difference between the technical and trade publishers' promotion practices lies with their respective publication schedules. Trade publishers normally center their publicity campaigns around the publication of their fall and spring "lists". Conversely, most technical publishers publish books year-round. A firm might publish as many titles in January (a dead sales month for a trade publisher) as it does in May or September. Finally, few technical books are bestsellers, at least in the popular meaning of the word. They are usually written by specialists *for* specialists, so that their appeal will be limited from the start. Hence, their need for radio and TV spots and full-page ads in the *New York Times Book Review* would be minimal.

"What difference does this make to me, the author?" you may ask. "There's not much I can do to promote my book. Anyway, that's the publisher's job." Indeed it is, but you are wrong in assuming that there is nothing you can do. You can do several things, none of which you are obligated to do, but all of which you *should* do, or at least try to.

PRIMING THE PUBLISHING PUMP

Recall that one article of your book contract probably said, in effect, that the publisher would send review copies of your book to selected periodicals and other media, and that he would follow the author's recommendations in this respect. These reviews are an excellent form of free advertising for the book, the publisher, and *you*.

Unfortunately, they aren't always easy to get. Of the thousands of technical books published every year, only a small fraction get reviewed. This is because many technical periodicals do not publish book reviews, and those that do only have space to run one or two feature reviews per issue. A "feature" review is one that discusses, at least generally, the book's contents and provides some critique of them. It should not be confused with a "book listing" that does no more than indicate the book's title, author, publisher, and price. The "Just Arrived" column in *Chemical Engineering* is an example of the latter. In each monthly issue, this magazine lists 20 or more new books, while printing only one feature review. That's only a dozen reviews each year. Other periodicals, such as the *Journal of the Air and Waste Management Association*, mention several new titles in each issue. Each title is summarized in two or three sentences — not a feature review, by any means, but certainly better than a perfunctory listing. Still other magazines (e.g., *Pollution Engineering*) run several reviews at a time — when they run them, which is sporadically. Most other technical periodicals fall within these limits re: book reviews.

What can you do to help get your book reviewed? First, make a list of the names, addresses, and book review editors of every periodical that could possibly be interested in reviewing your book. Next, glance through each of these periodicals to see what sort of review format it uses, how much space it allows each, whether the reviewers are staff or freelancers, what kinds of books it prefers to review, etc. Then, send this list to the publisher, along with your recommendations as to which periodicals should be targeted specifically. This is where personal contacts come in handy. Your publisher is undoubtedly well acquainted with several periodical editors and enjoys a nice "quid pro quo" relationship with them. He may be able to get your book valuable review space. But consider: he may publish scores of books every year. He can't expect to get special consideration for every one of his titles. This is where you come in. Perhaps you've written several articles for a certain periodical — articles that the editor asked you to write (for zero honoraria) and that were well received, to boot. To put it bluntly, he owes you a favor, and if he's reasonable, would be glad to give your book review space. Bear in mind, of course, that this doesn't guarantee your book a *good* review; it merely assures it *a* review.

Even if you have no inside track with an editor, you can still contact her to request that your book be reviewed. Sometimes, an editor will be more receptive to such a request coming from an author than from a publisher, especially if she finds the subject of the book intriguing, or if you've made it so.

A second way to help promote your book is by collecting names and addresses of potential readers — in industry, academia, government, or the consulting world. Better yet, try to locate mailing lists of potential readers. These include membership lists of professional organizations, trade associations, and other entities. Publishers can purchase or rent some of these lists for reasonable amounts. (However, some organizations refuse to divulge their lists to protect their members' privacy.) Other lists are available from reference books, such as the *Thomas Register* and *Peterson's Guide to Graduate Programs in Engineering and Applied Sciences.* (The latter lists the name, address, and head of every accredited graduate program for each of a variety of several engineering and scientific disciplines.)

The most efficient way to make use of these lists is to send them to the publisher. He is equipped to make mass mailings of flyers and other materials to advertise your book, while you aren't. Besides, with his bulk mailing permit, he can send an ad for only a fraction of the postage cost you'd have to pay. However, if you're willing to spend a little time and money, you can mail a limited number of flyers yourself, perhaps a few hundred. The advantage of this is that you can include with each flyer a "personal note" via which you could try to establish a one-to-one rapport with the addressee. This note needn't be elaborate. It only needs to highlight some of the book's features and how and why these features would be of interest to the addressee. Consider the following example:

Dear Mr./Ms./Dr._____:

President Quackenbush has just signed into law the Safe Highways Act of 1991. This Act contains a $50-billion appropriation for repair of our nation's superhighways and bridges. Highway engineering and engineering economics will play a very big part in implementing the provisions of this Act, especially in the next 20 years. My book, *THE U.S. INTERSTATE HIGHWAY SYSTEM: A Socio-Technical History*, should make it easier to do so, especially without repeating past road- and bridge-building mistakes. As the attached flyer explains, this book contains essential information for anyone — land-use planner, civil engineer, public administrator, or contractor — who is or will be involved with a highway or bridge repair project. To my knowledge, no other book provides such a unique, comprehensive perspective of these essential projects. Although it contains a great deal of up-to-date technical and cost information, *THE U.S. INTERSTATE HIGHWAY SYSTEM* is easy to use, logically organized, and oriented toward the solution of everyday problems. Moreover, the book would provide graduate or senior undergraduate engineering, planning, or public administration students with informative reading to challenge their analytical skills.

To order the book, please phone Median Strip Publishing at (800)-555-4321. If you have any questions once you receive your copy, please write or phone me (191-555-7711). I would be delighted to answer them.

Warmest regards,

Edward E. Edwards, P.E.

A personal note like this — especially if it's hand-addressed and signed — can get the addressee's attention more effectively than a packet of flyers stuffed in an envelope bearing a computer-generated mailing label. But again, the cost of mailing them (in both time and postage) can be significant.

A third way to publicize your book involves you even more personally; that is to travel to technical conferences, bookseller conventions, and like gatherings with your publisher. Here, you can make yourself available to sign copies of your book, answer questions about it, and generally provide the personal touch that, in my opinion, is sorely lacking in the book publishing world. Of course, if you're like most of us, you have other responsibilities that may preclude you from attending these functions. Besides which, there is the matter of your travel expenses. You couldn't expect your employer to pay for these trips, as they'd have little to do with company business. However, if you're willing to make the trips on your own time, and the publisher is willing to pay your way, these gatherings can be very rewarding.

Finally, you can interact directly with the publisher's marketing department in such areas as the design of ad flyers and the planning of mass mailings. Consider the former. For many technical books, ad flyers are the main way to reach potential buyers. Each flyer contains (or should contain) the following information: the title, author, and author's affiliation, all boldly displayed; a summary of the book's contents; a list of the kinds of specialists who would benefit from buying it; and information on ordering copies.

Unfortunately, most of the dozens of flyers that potential buyers regularly receive contain these data. The trick is to design the flyer for your book so that it stands out in the crowd. The marketing staff probably have certain standard flyer configurations that they use, and they may not be too receptive to your creative suggestions. But, it doesn't hurt to ask. Another way to spruce up your flyer is to include quotations from book reviews (favorable ones, naturally!) and testimonials from "satisfied readers". Re: the latter, such statements as "I find it to be an excellent and much-needed reference" are valuable, as well as eye-catching, especially if the person quoted is well known in the field. How do you obtain these testimonials? Sometimes, readers will write you directly to express their pleasure with the book. With their permission, you can use portions of their letters. Other times you may have to solicit them directly.* Ask your publisher to send copies of the book to your associates, leading professionals in your field, and anyone else that might respond favorably to it. You have little to lose and much to gain.

Well, we've covered a lot of ground in this very long chapter. By now, you should have a fairly good idea of how to: select a publisher to query, structure your query, deal with your editor, write your book, and promote it once it's published. In the next chapter, we'll take a closer (and more critical) look at the technical writing market.

---

* Although this may seem like a presumptuous thing to do, it's done all the time in the trade publishing world. The next time you pick up a bestseller, especially a paperback edition, note the quotations from famous persons splashed all over the covers. In most cases, these testimonials were not unsolicited. The celebrities were sent copies of the page proofs and, after reading same, returned their comments. Needless to say, only the most favorable of these were used.

REFERENCES

1. Bly, R. W. *Secrets of a Freelance Writer: How to Make $85,000 a Year* (New York: Dodd, Mead, and Company, 1988).
2. Collier, O. *How to Write and Sell Your First Nonfiction Book* (New York: St. Martin's Press, 1990).
3. Tennant Neff, G. (Ed.) *1991 Writer's Market* (Cincinnati: Writer's Digest Books, 1990).

5   The Technical Publishing Market: An Overview

*The first person a writer has to sell is not an editor
or a publisher, but himself.*

Roy Meador, letter to the author

If sheer numbers are any indication, the technical writing market is very healthy these days. According to standard publishing references, there are thousands of publishers and tens of thousands of periodicals in business today. Many of these are technical publishers and periodicals sorely in need of good material. The subjects that interest them cover the universe of scholarship. In fact, it'd be difficult to name a field, no matter how esoteric, that has not been the subject of a book or a periodical. For anyone interested in technical publication, the market would seem to be promising indeed.

Despite these statistics, getting published is rarely an easy endeavor, especially by the leading periodicals and publishers. The competition can be quite keen, regardless of the size of royalties/honoraria paid. In fact, the competition arises more from the reputation-enhancing rewards of publication than it does from the financial rewards. This is fortunate, as technical writing, especially for periodicals, is *not* very rewarding financially. (If the surveys I made are in any way indicative, most periodicals pay nothing, while those that do pay modestly.) Technical book writing does offer better financial returns than writing for periodicals. However, the returns to the author depend upon the number of copies a book sells. If a book sells well (say, 1000 copies or more), an author can realize a few thousand dollars. (I'll discuss the results of these book and periodical surveys later in this chapter.)

However, money is not the only consideration, of course. To many, the act of writing, the sheer joy of self-expression, is pleasure enough to justify the hard work that it entails. Seeing one's work in print in a prestigious journal or on the cover of a book affords an added thrill. Add to this the professional benefits that publication can bring and you have a very sound "rationale for writing".

THE BOOKS

In Chapter 4, I described, at length, the several hurdles you have to clear to get your book published. From querying editors to autographing flyleafs, the process is long, laborious, and often frustrating. Yet, a book provides the most permanent record of your work. Or, as a senior editor at a leading engineering journal once told me, "When you write a book, you're building a monument to yourself."

But every monument has to have a pedestal and someone to set it thereon. That's a where a publisher comes in. It would be very difficult for you to get your book printed without a publisher.* Publishers know how to produce, print, and market books, while earning their firms respectable profits. Few technical professionals have this experience.

Who are the technical publishers, where are they located, who directs them, and what are their terms? Answers to a few of these questions can be found in publishing industry references, though only the publishers themselves can answer all of them. As I mentioned in Chapter 4, among the two most commonly used and cited references are the *Writer's Market* and the *Literary Market Place* (LMP). Updated annually and moderately priced (under $25), the *Writer's Market* lists market information for most of the trade (mass market) American publishers and a few of the larger technical publishers (e.g., McGraw-Hill). The data, however, are usually limited to such statistics as the number of books published annually. No sales figures are provided. For some publishers, *Writer's Market* lists the number of proposals received annually. This statistic is both a measure of the publisher's size and its popularity among writers. Generally, the larger the publisher, the greater the number of submissions. As a writer, you would be interested in the "P/P ratio" — the number of proposals divided by the number of books published annually. Usually, the higher this ratio, the lower the odds of getting them to publish your book. For instance, in the *1991 Writer's Market*, one publisher of consumer-oriented books said that it received

---

* Not that it hasn't been done. A number of technical books have been self-published; for example, Levenspiel's *The Chemical Reactor Minibook* referred to in Chapter 3. However, self-publishing is an expensive, time-consuming endeavor that a writer shouldn't attempt, unless he possesses business acumen, a Rolodex full of contacts, and more than a smidgen of luck.

1000 proposals every year, but only published 30 to 35 books. That's a P/P ratio of about 30 to 1.[1]

Subtitled *The Directory of the American Book Publishing Industry*, the LMP lists 2691 U.S. and Canadian publishers, making it the most comprehensive reference of its kind.[2] The LMP also provides information on other aspects of the publishing industry, including editorial services and agents, marketing and publicity, book manufacturing, and sales and distribution. Each publisher entry contains the following information:

- Full name, mailing address, and telephone number
- Name and title of chief executive
- Brief description of kinds of books published
- Numbers of titles published and in print in 1990
- Year when founded
- ISBN prefix(es) (first several digits of the unique number given to each published book)

Entries for the larger firms also contain such data as names of senior staff (executives, editors, marketing directors, etc.), subsidiary publishers, and foreign representatives. Though the LMP contains no market-oriented statistics, it is nonetheless valuable as *the* publishing industry directory.

A book review column in a technical periodical offers another, albeit limited, information source. Each book review always lists the publisher's name and, in some periodicals, his address as well. With these data in hand, consult the LMP to obtain the publisher's phone number, names of senior staff, and other relevant information.

As invaluable as the *Writer's Market* and the LMP are as publishing industry information sources, they provide just the bare minimum. But, as a writer, you probably would like to know more about technical publishers, especially the one for whom you might be writing a book. Such information as how many books the firm publishes annually and the number of annual submissions would be useful, as would the royalty rate you would receive and the number of copies you could expect your book to sell. Naturally, these statistics will vary from publisher to publisher, let alone book to book. Nonetheless, by surveying a number of technical publishers, we could obtain information that is at least illustrative, if not definitive.

With just this objective in mind, I sent a questionnaire (Figure 5.1) to about 2 dozen technical publishers. Of those that responded, most provided only the most basic information — name, address, phone number — data that could be obtained from LMP or some other reference. They did not reply to the other questions, saying either that they did not have the information or that they could not provide it for reasons of

Please answer the following questions as completely as youcan and return this form in the enclosed S.A.S.E. Thank you very much!

- Editor-in-Chief: _____
- Phone no.:_____
- Fax no.:_____
- Subjects of books published (e.g., medicine):_____
- No. of titles published annually: _____
- Approx. no. of annual submissions:_____
- Please provide the following data for a "typical" book:

    — Production time (months):
    — Royalty schedule:
      * ___% on first _____ copies sold
      * ___% on next _____ copies sold
      * ___% on copies sold over _____
    —Terms on foreign and other sales: ____ % of base royalty
    —Copyright is owned by: ____ (publisher) ____ (author)
    —Advance against royalties (if any): $ _____
    —List price: $ _____
    —Approximate number of pages (total): _____
    —Illustrations? _____ (yes); _____ (no)
    —First printing: _____ copies
    —Additional printings (number): _____
    —Copies sold in first year: _____
    —Copies sold in second year: _____
    —Copies sold in years 3 to 5: _____
    —Years book is in print: _____
- Proposal specifications:
    — Query letter? ____ (yes) ____ (no)
    — Outline and preface?: _____ (yes) ____ (no)
    — Other? (please specify):
- Manuscript specifications:
    — Double-spaced?: _____ (yes) _____ (no)
    —Letter quality required?: _____ (yes) _____ (no)
    —Computer diskette required?: _____ (yes) _____ (no)
- Suggestions, comments, etc.:

Figure 5.1. Publisher questionnaire.

confidentiality. However, a few publishers *did* reply fully to the questionnaire. Their responses are summarized in Table 5.1. (Note: some of the publishers allowed me to use their responses only if I promised not to divulge their names.) Although the others did not impose this condition, I decided, for consistency's sake, not to name any of them. Referring to Table 5.1, notice that many data categories have two entries. The first is an average of the responses received. The second entry,

Table 5.1. Technical Book Publisher Questionnaire Responses

Annual production figures
    Titles published: 450 (95–1991)
    Submissions: 520 (200–900)
Statistics for a "typical" book
    Production time (months): 7 (6–9)
    Royalty schedule:[a]
        10% on first 1000 copies sold
        12% on next 2000 copies sold
        15% on copies sold over 3,000
    Terms on foreign or other sales: 50% of base royalty[b]
    Copyright ownership: publisher[b]
    Advance against royalties: "negotiable"; "determined by book"
    List price: $75 ($49.95–$90)
    Length (pages): 400 (200–800)
    Illustrations included?: yes
    First printing (copies): 1800 (1000–4000)
    Additional printings (number): 1 (1–2)
    Copies sold (per year):
        First year: 900 (200–1500)
        Second year: 500 (200–1000)
        Third to fifth year: 250 (100–500)
    Years in print: 6 (3–10)
Submittal specifications
    Proposals: query letter *plus* outline and preface
    Manuscripts:[c]
        Double-spaced
        Letter quality? not required, but preferred
        Computer diskette required with hard copy? yes (50%); no (50%)

[a] For some respondents, this was a "negotiable" item. However, none offered less than a 10% royalty.
[b] These appear to be "boiler plate" terms in all technical publishing contracts.
[c] The requirements of respondents varied in this area. Some were indifferent to letter-quality typescript and computer diskettes, while others required both with manuscript submittals.

in parentheses, shows the overall range in the data provided for the category. The following example will illustrate how these entries were developed:

Example. Five publishers gave the following responses for a category: 200, 400, 750, 800, and 200–900, in turn. To determine the average, we first calculate the average response from the fifth publisher; that is, (200 + 900)/2 = 550. Then, using this average response and the other four responses, we compute the average for the publishers: (200 + 400 + 750

+ 800 + 550)/5 = 540. Both this average and the overall range would be reported in the table, i.e., "540 (200 – 900)".

In some cases, the range (e.g., number of titles published annually) is broad, while in others (e.g., production time) it is narrow. For the "advance against royalties" category, respondents gave no numbers, just such responses as "negotiable" or "determined by book". Strictly speaking, those responses are honest ones. Yet, one cannot help wondering whether these firms were perhaps reluctant to discuss what is, both to writers and publishers, a sensitive topic.

Apart from that, what do the Table 5.1 data tell us? To answer that, let's look at each category individually.

Annual Production Figures

The "titles published" and "submissions" averages are approximately equal. But as the ranges in parentheses indicate, this comparability is somewhat illusory. The "titles published" number covers a 20-fold range, while the "submissions" range is much narrower. How can a publisher publish more books than he receives requests to publish? Very readily, in fact. For one thing, firms often publish certain books, such as references, in successive editions, some appearing as often as annually. A publisher need not receive a new submission from an author to publish a new edition. He would merely ask the author(s) to update the book with new material. Also, some books published are staff-written, meaning that *no* submissions have ever been prepared for them. These considerations aside, the responses do imply that the odds of a book proposal being accepted are still relatively high, at least higher than with a typical trade publisher. (The 30:1 proposals:books published ratio discussed above is more typical than atypical.)

The next part of the questionnaire addressed the "typical book". That is, each respondent was asked to provide certain statistics on a hypothetical book. This "typical" book would be a composite of the books that he publishes. Like the "typical American family", this book probably doesn't exist, but it's quite useful to have as a model of the real technical books that do get published.

Production Time

On average, it would take 7 months to edit, typeset, and print a typical book. This time fell within the relatively narrow range of 6 to 9 months reported by respondents. To a research scientist used to working on a project for years before seeing any usable results, 7 months might seem like a short time to produce a book. Conversely, to a maintenance engineer responsible for repairing or replacing dozens of pieces of process equipment during a 2-week plant maintenance shutdown, 7 months might seem like an eternity. In any event, this production time is not only typical, but is also somewhat less than the time required to produce a book before the advent of computerized typesetting. In those days, the production time was more like 9 to 12 months.

Royalty Schedule

There was little variability here. As Table 5.1 indicates, the typical schedule is 10% on the first 1000 copies sold; 12% on the next 2000; and 15% on sales exceeding 3000. For some respondents, the royalty schedule was "negotiable". None of the publishers offered less than a 10% royalty rate, although some paid no more than 10%, regardless of the number of copies sold.

Terms on Foreign or Other Sales

As I said in Chapter 4, the royalties paid an author are based on the revenues generated by the book. And these revenues are, in turn, determined by the price(s) at which the various copies are sold. This would seem to be a fair arrangement. Unfortunately, nearly every technical publisher's contract contains a clause stating, in effect, that a *lower* royalty rate is applied to foreign or other sales where the copies are sold at prices less than some fraction (typically, 75%) of the list price. This lower rate is usually one half of the base royalty rate, which is either 10%, 12%, or 15%, depending on the total number of copies sold. One reason for including this clause is that the publishers' revenues on these sales are less than what he would realize from domestic sales, due to his having to pay overseas firms a portion of his revenues for distributing the book. Another reason — one that's easier to understand, though perhaps not easier to accept — is that the publisher's profit margin on these heavily discounted sales is lower. Consider this simple example: if the cost of producing a $50 (list price) book were $20, the publisher's profit (before royalty payments) on a book sold at the list price would be $30 ($50 – $20). Let's assume that the author's royalty on that book would be 10% of $50, or $5. However, if the publisher sold that book at 70% of list (i.e., $35), his profit would be halved, to $15. The author would normally expect to receive $3.50 from the discount sale, but since the book was sold at less than 75% of list, he would only get a 5% royalty (one half of 10%) or $1.75.

Is this fair? From the publisher's perspective it is, because large discounts cut into his per-book profit margin, and he has less of a fund (if you will) from which he can pay royalties. Besides, the publisher reasons that discounting boosts sales, increasing both revenues and royalties, accordingly.

Authors might agree — but only up to a point. They might reason that the publisher receives the lion's share of the revenues anyway, so that he should be willing to at least pay the same royalty rate on *all* sales, regardless of the price at which the books are sold. Moreover, authors might ask, who wrote the books in the first place?

Clearly, there are two sides to this issue. Perhaps the root of the matter lies in the fact that, in technical and trade publishing, the royalty rate is based on sales *revenues*, rather than on sales *profits*. If publishers paid royalties on profits (higher rates, of course!), this issue could be avoided altogether. However, that would complicate the publisher's bookkeeping, while not necessarily increasing the author's royalties. Be that as it may, the fact is that most technical book contracts contain this clause; and to my knowledge, it is *not* negotiable.

Copyright Ownership

For our typical book, the copyright is in the publisher's name. As discussed in Chapter 3, it makes little difference who owns the copyright to a technical book. Subsidiary sales rarely arise; when they do, they don't amount to much. The exception is translations, and they are usually covered in the contract. In any event, copyright ownership is normally a "boiler plate" contract clause.

Advance Against Royalties

As shown, this is typically a "negotiable" item. However, the subject of advances won't be discussed, unless the author raises it. Unlike a trade book publisher, who almost always offers an author an advance (albeit a modest one) at the beginning of contract negotiations, a technical publisher usually will not. This is not to say that a publisher will refuse to discuss the matter. Quite the contrary. If the author requests a reasonable advance, the publisher may very well grant it. How much is "reasonable?" An advance equal to the projected first-year's royalties on the book seems to be a fair amount.

List Price

Compared to trade hardcover books, technical books are pricey. Their higher prices are primarily due to lower sales. A trade book that sells 20,000 copies can be sold for much less than a technical book that sells one tenth as many. In addition, technical books are usually made with higher quality paper and bindings, contain more illustrations, and consist of many more equations and other hard-to-typeset material. However, as with trade books, the list price is somewhat deceptive, as most books are sold for less. The range shown in Table 1 ($49.95 to $90) covers most technical titles, but not all. I know of books listed in the $25 to $30 range, and others priced near $150. Nonetheless, $75 is "typical".

Length

The price is also determined, in part, by the length, as the length partially determines the typesetting and printing costs. The average reported value of 400 pages (including appendices and indices) falls in the lower half of the 200- to 800-page range. Again, I'm familiar with both shorter and longer books, but these are atypical. Any book shorter than 200 pages is hardly deserving of the name, let alone the price.* But there are exceptions: Roy Meador's highly successful *Guidelines for Preparing Proposals* runs to a whopping 123 pages, including front matter. Conversely, a book longer than 800 pages is probably too long for a monograph, but not so for a handbook. Indeed, books longer than this are almost always handbooks. Again, there are exceptions.

Illustrations

Unsurprisingly, a typical technical book includes illustrations, be they photographs, line drawings, or any of a wide range of graphical images. In many ways, this is what separates technical books from trade and other kinds of titles.

First Printing

The average response for this category was 1800 copies. The range (1000 to 4000) reflects the variability among publishers as well as books. For instance, a prominent publisher with an extensive marketing network and a large warehouse capacity might order a first printing of 2000 copies of a book of which a smaller publisher might order only 1000. The book makes a difference, too. The initial printing of the 5th edition of a popular textbook might be more than 4000 copies, while that of a first edition monograph might be only 500 copies. Interestingly, the trend among publishers is to order larger first printings, mainly because printers offer much lower per-unit prices on larger print runs.

Additional Printings

If a book sells well enough, a publisher may order a second printing. When and if he does so will depend on the monthly sales figures, averaged over several periods, as well as the number of copies left in stock. If the sales show a consistent upward trend and are so high that the inventory would be exhausted in 2 or 3 months, a second printing would probably be ordered. However, if the sales have been low or erratic since publication, the book may never see a second printing; but, on the average, a technical book is reprinted once, at least according to the questionnaire respondents.

---

* This title excepted, of course!

Copies Sold

As one might expect, a book sells best during the first year of its "life". The responses bear this out. During the first year, a typical book sells, on average, 900 copies; during year two, 500; and during each of years 3 to 5, 250 copies. Thus, over the 5-year period, it sells a total of 2150 copies — just over one printing's worth. But notice how much the responses varied. For instance, the first year's sales figures exhibited a sevenfold range. Given the wide variability in sales, these responses should not surprise us.

Years in Print

A typical book is in print 6 years, though it could be in the publisher's catalog as little as 3 or as many as 10 years. Obviously, its sales record will be the main determinant of this. So will the number of copies left in the warehouse. If a publisher has recently ordered a 2000-copy second printing of a top-selling book that has, inexplicably, cooled off, he will likely keep selling it until most of those copies are gone; and that could take years. To move some of this backlog, the publisher could "remainder" the books; that is, sell them at a big discount to bookstores or book clubs. However, many publishers prefer not to, as remaindering cuts deeply into their revenues, not to say the authors' royalties. From the publisher's standpoint, it is better to keep selling the book at (or near) the list price. This makes good business sense, especially if the information in the book continues to be up-to-date.

Submittal Specifications

Before they will agree to publish a given book, most technical publishers want to see a proposal first. This proposal is typically a preface and outline of the proposed book. As discussed in Chapter 4, the outline may be a table of contents, listing of main topics, or any other format that concisely conveys what the book will be about. Of course, to properly "introduce" this material — and himself — to the publisher, the author should also enclose a query letter. On rare occasions, the author may have already written the book and is looking for someone to publish it. In this case, a publisher would probably like to see the entire manuscript, but only after being asked first, again via a query letter.

As to manuscripts, most publishers are fairly liberal in their requirements. They only ask that they be legibly typewritten, double-spaced, on $8^1/_2 \times 11$-inch paper (preferably white). Of those surveyed, none required "letter quality" manuscripts, though they preferred them. However, half of the respondents require that authors submit the manuscript in both hard copy and computer diskette forms. The latter is required for two reasons: (1) it is much easier to edit a diskette on-screen than hard copy that would have to be retyped, and (2) most typesetting processes nowadays are computerized, using diskettes as the "input".

"TYPICAL BOOK" REVENUES AND ROYALTIES

If we wish, we can use the above information to predict the revenues and royalties generated by sales of our typical book. These predictions are listed below. They are based on the data in Table 5.1 and an assumed average sales price of $60 — 80% of the $75 list price.

	Revenues	
Copies sold	To publisher	To author
First 1000	$54,000	$6,000
Next 1150	60,720	8,280
Totals	$114,720	$14,280

According to Table 5.1, nearly one half (900) of the total copies sold (2150) will be sold in the first year. However, because the royalty rate on that amount is just 10%, the author's royalties during the first year will be proportionately lower, or $5400 (900/1000 × $6000). Though not a king's ransom, this is nonetheless a respectable return. Moreover, it would be a respectable (and reasonable) advance against royalties. During the years the book is sold, the author's share would be $14,280, approximately 11% of the total sales revenues. Averaged over the assumed 6-year "life" of the book, this would be about $2400 per year.

THE PERIODICALS

As explained in Chapter 4, the approach to getting a paper or article published is somewhat different from that in getting a book in print. For one thing, the numbers are different. Counting foreign serials, there are 40 times as many periodicals as there are book publishers. *Ulrich's International Periodicals Directory 1990–91* lists 116,000 serials that are issued "at regular intervals (i.e., monthly), as well as those issued annually or less frequently...or irregularly."[3] In fact, Ulrich's includes "all publications that meet the definition of a serial except general daily newspapers, newspapers of local scope or local interest, administrative publications of major government agencies below state level that can be easily found elsewhere, membership directories, comic books, and puzzle and game books." All well and good, but how many of these are technical periodicals? Ulrich's does not separate their entries into "technical" and "nontechnical" categories. However, it does identify referee serials because they "represent the core of research materials, which is one of the reasons we have chosen to highlight those publications which pass peer review scrutiny." Nearly 2500 refereed serials are identified. Thus, *at least* this number of technical periodicals are being published today. Due to the large number of serials and the sheer diversity of subject matter published, it is more difficult to characterize the periodicals market than it is the book business.

Each of the thousands of entries contains basic information about the periodical; namely (1) main entry title, (2) frequency of publication, (3) publisher address, (4) country code, and (5) Dewey Decimal Classification number. Additional information is reported for some periodicals, including:

- Year first published
- Price
- Subscription or distribution address
- Telephone, fax, and telex numbers
- Editor's name
- Special features (e.g., book review column)
- Circulation
- Format (nonstandard, such as loose-leaf)
- Abstracting and indexing
- Online/CD-ROM availability
- Description of contents and editorial focus
- Refereed serial designation (if applicable)

Though the above information is of interest to an author, none of it addresses such matters as manuscript submittal specifications or payment terms. The *1991 Writer's Market* (WM) does include fairly complete market data, but only for a handful of technical periodicals. (Most of the WM entries pertain to the trade periodicals from which professional writers earn their bread.) In an attempt to bridge these "data gaps," and at the same time, to develop a profile of the market, I sent a questionnaire to 25 technical periodicals. (See Figure 5.2.) I received usable replies from 19 periodicals. The results are shown in Appendix 5.1.

The responses cover a range of contents, submittal specifications, and payment terms. Nonetheless, some commonalities emerge:

- Few of the technical periodicals pay their contributors. This is especially true of the refereed journals, the ones generally the most difficult in which to get published. Furthermore, most of the articles that are purchased are news features. The rates for these range from roughly $0.25 to $0.40 per word to a flat fee of $350.

Please answer the following questions as completely as you can and return this form in the enclosed S.A.S.E. Thank you very much!

- General information:
    Periodical title: _____
    Address: _____
    _____
    Editor-in-Chief: _____
    Phone no.: _____
    Fax no.: _____
    Circulation: _____
    Frequency of publication: _____ (monthly) ____ (bimonthly) _____ (quarterly)
    _____ (other)
    Refereed (i.e., peer-reviewed)? ____ (yes) ____ (no)

- Contents:
    Please briefly describe the subjects typically covered by articles published in your journal (e.g., environmental issues). Also, please indicate the types of articles that appear (i.e., features, reviews, interviews, technical notes, etc.) and how many of each you purchase from outside authors in a typical year.

- Submittal specifications:
    Query letter or entire manuscript? _____ (Q) _____ (MS)
    Length: _____ words
    Letter quality required? ____ (yes) ____ (no)
    S.A.S.E.? _____ (yes) _____ (no)
    Time required to respond to query: _____ weeks
    Time between acceptance and publication: _____ months
    Photos/illustrations preferred? _____ (yes) _____ (no)

- Terms:
    Payment: _____ cents/word or $ _____ (fee/honorarium)
    Kill fee (if any): _____% of base rate
    Rights bought: _____ (all) _____ (1st serial only)
    When paid: _____ (on acceptance) _____ (on publication)
    Byline given? _____ (yes) _____ (no)
    Pay for photos/illustration: $ _____ /item

- Suggestions, comments, etc.:

Figure 5.2. Periodical questionnaire.

- Those periodicals that do pay for writing purchase *all* the publishing rights to a piece. That is, the author is not permitted to reprint the article in another journal or anywhere else without first getting the periodical's permission. By contrast, most sales to trade magazines are for the "first North American serial rights" only. This just gives the periodical the right to publish the piece one time. Indeed, according to the 1976 Copyright Act, the author automatically retains all other rights to the piece, unless he or she agrees to relinquish them. (We'll discuss copyright more in Chapter 6.)

- The time between acceptance of a manuscript and its publication is usually several months, 4 to 6 being typical. This delay reflects the time needed for peer review and the fact that most periodicals work several issues ahead. It may also reflect a manuscript backlog. However, most periodicals respond to article queries within 4 weeks so that authors are not left in the lurch. (Conversely, trade periodicals often take months to reply.)

- Editors of technical periodicals generally do *not* require that a S.A.S.E. (self-addressed, stamped envelope) be enclosed with a query. Nevertheless, a submitter should enclose one as a courtesy. It costs so little to do so.

- It is also courteous to submit a letter-quality manuscript (or equivalent), even if the periodical doesn't require it (and most of those listed in Table 2 don't). Dot matrix printer submissions are often difficult to read, unless the printer has been set to "near letter quality" (NLQ) mode. (NLQ manuscripts are almost as easy to read as letter quality.) Keep this maxim in mind: "readable manuscripts are most salable."

- A final observation: although the circulation — and readership — of these periodicals varies over a wide range (roughly 3000 to 50,000), none can be said to have a large circulation, at least by trade magazine standards. However, remember that these periodicals, like most technical journals, are targeted to specialists. Thus, proportionately speaking, the chances of an article being read by one of these specialists are much greater than the odds of someone reading a piece that appears in a trade magazine. The impact of such an article would be proportionately greater. In the final analysis then, what counts most isn't circulation, but readership.

Albeit brief, this look at today's technical writing market does shed some light on a world that, to many technical people, has long been shadowy. To repeat, this information is far from comprehensive. A thorough study of the market (i.e., an expansion of Tables 1 and 2) would require hundreds of pages. For you, a prospective author, the best way to learn about the writing market is not to study publishing references or even this chapter, but to investigate the marketplace personally. That means getting acquainted with the editors of the leading technical periodicals in your field. After all, first-hand information is still the best.

REFERENCES

1.   Neff, G. T. (Ed.) *1991 Writer's Market* (Cincinnati: Writer's Digest Books, 1990).
2.   *Literary Market Place 1991* (New York: R.R. Bowker, 1990).
3.   *Ulrich's International Periodicals Directory 1990–91* (New York: R.R. Bowker, 1990).

Appendix 5.1

MARKET INFORMATION FOR SELECTED TECHNICAL PERIODICALS

1. *American Industrial Hygiene Association Journal*

   General information:
   Address: P.O. Box 8390, 345 White Pond Drive Akron, OH 44320
   Editor-in-Chief: Paul D. Halley
   Phone: (216)-873-2442
   Fax: (216)-873-1642
   Circulation: 11,500
   Frequency of publication: monthly
   Refereed (peer-reviewed)? yes

   Contents:
   "Health and safety issues concerning the workplace; industrial hygiene; environmental issues in health and safety. Every month we publish the latest research in industrial hygiene and related fields, the latest products available, a calendar of meetings and courses, book reviews, and features by our regular and guest columnists."

   Submittal specifications:
   Query letter or complete manuscript? manuscript
   Letter quality required? yes
   Self-addressed, stamped envelope required? no
   Query/manuscript response time: 1 week
   Time between acceptance and publication: 4 months
   Photos/illustrations preferred? yes

   Terms:
   In most cases, this periodical does not pay contributors. Exception: "The only material we purchase comes from our regular columnists."

2. *AIChE Journal*

General information:
    Address: 345 East 47th Street New York, NY 10017
    Editor-in-Chief: Mark D. Rosenzweig
    Phone: (212)-705-7576
    Fax: (212)-752-3294
    Circulation: 4000 (approximately)
    Frequency of publication: monthly
    Refereed (peer-reviewed)? yes

Contents:
    "Important research work in the field of chemical engineering; review articles of chemical engineering emerging technology; technical notes."

Submittal specifications:
    Query letter or complete manuscript? manuscript
    Letter quality required? no
    Self-addressed, stamped envelope required? no
    Query/manuscript response time: 2 to 4 weeks
    Time between acceptance and publication: 4 to 6 months
    Photos/illustrations preferred? yes

Terms:
    This periodical does not pay contributors.

3. *American Journal of Public Health*

General information:
    Address: 1015 15th St., N.W. Washington, D.C. 20005
    Editor-in-Chief: Michel A. Ibrahim, M.D., Ph.D.
    Phone: (202)-789-5600
    Circulation: 35,000
    Frequency of publication: monthly
    Refereed (peer-reviewed)? yes
Contents:
    According to its "Information for Authors," this periodical publishes a variety of material: "'Articles', no longer than 5000 words, or 'Public Health Briefs', no longer than 1000 words, report original research, unusual outbreaks of disease, program evaluation or new methodologies of broad interest, reviews that are analytic rather than descriptive in nature. 'Commentaries', no longer than 5000 words, deal with policy analysis in an analytic fashion. 'Different views' presents two or more papers with contrasting viewpoints regarding data analysis or opinion. 'Letters', no longer than 400 words, may critique material the Journal has published or present information of general interest. 'Editorials' take off from a paper published in the same issue of the Journal and are solicited on an individual basis by the editor."

Submittal specifications:
    Query letter or complete manuscript? manuscript
    Letter quality required? yes
    Self-addressed, stamped envelope required? no
    Query/manuscript response time: varies
    Time between acceptance and publication: 6 months

Terms:
    This periodical does not pay contributors.

4. *Biotechnology Progress*

General information:
    Address: University of Pittsburgh 911 William Pitt Union Pittsburgh, PA 15260
    Editor: Jerome Schultz
    Phone: (412)-648-7956
    Fax: (412)-624-7145
    Circulation: 3000 (approximately)
    Frequency of publication: bimonthly
    Refereed (peer-reviewed)? yes

Contents:
    "Significant original research related to biotechnology; review articles; technical notes."

Submittal specifications:
    Query letter or complete manuscript? manuscript
    Letter quality required? no
    Self-addressed, stamped envelope required? no
    Query/manuscript response time: 2 to 4 weeks
    Time between acceptance and publication: 4 to 6 months

Terms:
    This periodical does not pay contributors.

5. *Chemical Engineering*

General information:
    Address: 1221 Avenue of the Americas New York, NY 10020
    Editor-in-Chief: Richard J. Zanetti
    Phone: (212)-512-2849
    Fax: (212)-512-4762
    Frequency of publication: monthly
    Refereed (peer-reviewed)? no

Contents:
    "News articles and outside-authored engineering practice articles, dealing with practical technological developments and practices of interest and use to engineers in the chemical process industries."

Submittal specifications:
    Query letter or complete manuscript? manuscript
    Length: variable
    Letter quality required? no
    Self-addressed, stamped envelope required? no
    Query/manuscript response time: 6 to 8 weeks
    Time between acceptance and publication: 2 to 8 months
    Photos/illustrations preferred? yes

Terms:
    Generally, this periodical does not pay contributors. However, they do purchase approximately 100 manuscripts from outside authors each year. Payments for the latter are negotiable.

6. *Chemical Engineering Progress*

General information:
    Address: 345 East 47th Street New York, NY 10017
    Editor-in-Chief: Mark D. Rosenzweig
    Phone: (212)-705-7576
    Fax: (212)-752-3294
    Circulation: 50,000
    Frequency of publication: monthly
    Refereed (peer-reviewed)? no

Contents:
    "Every issue: at least one article on each of the following: energy transfer/conversion; environmental protection; fluids/solids handling; materials; measurement and control; safety; reactions and separations. All are features; no more than six are purchased per year. All are outside-authored, though."

Submittal specifications:
    Query letter or complete manuscript? both, though query preferred
    Length: no limit
    Letter quality required? no
    Self-addressed, stamped envelope required? no
    Query/manuscript response time: 3 to 4 weeks
    Time between acceptance and publication: 3 to 6 months
    Photos/illustrations preferred? yes
    Other: "Upon acceptance of a draft, we request a floppy-disk copy of the paper (in $3^1/_2$" MacIntosh format, preferably), as well as a hard copy."

Terms:
    This periodical does not pay contributors.

7. *Chemical Week*

General information:
Address: 810 Seventh Avenue, Ninth Floor New York, NY 10019
Editor-in-Chief: Peter Coombes
Phone: (212)-586-3430
Fax: (212)-586-3218 or -3147
Circulation: 48,000
Frequency of publication: weekly
Refereed (peer-reviewed)? no

Contents:
"News and features covering the international chemical industry. Includes any aspect of business, from sales and earnings, mergers, acquisitions, new projects, market reports, company and people stories, to environmental, science, and technology issues."

Submittal specifications:
Vary from project to project.

Terms:
$12/column-inch. (This is an average rate; actual rate is negotiable.)

8. *Coal*

General information:
Address: 29 North Wacker Drive Chicago, IL 60606-3298
Editor-in-Chief: Mark Sprouls
Phone: (312)-726-2802
Fax: (312)-726-2574
Circulation: 25,000
Frequency of publication: monthly
Refereed (peer-reviewed)? no

Contents:
"Coal mining technology; features (including mine case histories); industry news; people; new products."

Submittal specifications:
Query letter or complete manuscript? query
Length: varies
Letter quality required? no
Query/manuscript response time: 2 weeks
Time between acceptance and publication: 3 months
Photos/illustrations preferred? yes

Terms:
Payment: $11/column-inch for news; variable, for features, photos/
    illustrations
Kill fee: none

Rights purchased: all
When paid? on publication
Byline given? features, yes; news, no

9. *Environment*

General information:
Address: 4000 Albemarle Street, NW Washington, DC 20016
Managing Editor: Barbara T. Richman
Phone: (202)-362-6445
Circulation: 16,000 (approximately)
Frequency of publication: 10 times per year
Refereed (peer-reviewed)? yes

Contents:
   "Subjects: all subjects in environmental science and policy. [It is] critical that the topics be treated objectively, intelligently, and authoritatively. All manuscripts are reviewed by our board of editors. Types of articles: main (≤5000 words); overview (≤2000 words); books of note; report reviews; 'Spectrum' (short summaries of interesting developments)."

Submittal specifications:
Query letter or complete manuscript? both
Length: see "Contents"
Letter quality required? yes
Self-addressed, stamped envelope required? "doesn't matter"
Query/manuscript response time: 8 weeks
Time between acceptance and publication: 2 to 4 months
Photos/illustrations preferred? yes

Terms:
Payment: "We rarely pay for main articles; [we] sometimes pay for 'Overview' and report reviews." However, they do pay for photos, the amount depending on their size or whether they're in color or black-and-white.
Rights bought: all ("We ask authors to sign [a] copyright release form.")

10. *Environmental Progress*

General information:
Address: 345 East 47th Street New York, NY 10017
Editor-in-Chief: Mark D. Rosenzweig
Phone: (212)-705-7576
Fax: (212)-752-3294
Circulation: 5,000
Frequency of publication: quarterly
Refereed (peer-reviewed)? yes

Contents:

"Practical articles related to the control of air, water, and solid waste pollution."

Submittal specifications:

Query letter or complete manuscript? manuscript
Letter quality required? no
Self-addressed, stamped envelope required? no
Query/manuscript response time: 2 to 4 weeks
Time between acceptance and publication: 4 to 6 months
Photos/illustrations preferred? yes

Terms:

This periodical does not pay contributors.

11. *Journal of the Air and Waste Management Association*

General information:

Address: P.O. Box 2861 Pittsburgh, PA 15230
Editor-in-Chief: Harold M. Englund
Phone: (412)-232-3444
Fax: (412)-232-3450
Circulation: 13,000
Frequency of publication: monthly
Refereed (peer-reviewed)? yes

Contents:

The subjects covered are: air pollution control technology, air pollution sources, effects, meteorology, modeling, monitoring, waste management (hazardous, medical, and municipal solid wastes), global environmental issues. Types of articles published are features (general interest), technical papers, and "Technical Notes."

Submittal specifications:

Query letter or complete manuscript? manuscript
Length: 4000 to 8000 words
Letter quality required? no ("draft is ok")
Self-addressed, stamped envelope required? no
Query/manuscript response time: varies
Time between acceptance and publication: 2 to 3 months
Photos/illustrations preferred? yes ("if relevant to ms")

Terms:

Payment: In general, this periodical does not pay contributors. However, they purchase approximately three features per year, at negotiated prices.
Rights bought: all
When paid? on publication
Pay for photos/illustrations: "rarely"

12. *Journal of the American Medical Association*

General information:
    Address: 515 North State St. Chicago, IL 60610
    Editor-in-Chief: George Lundberg, M.D.
    Circulation: 390,000 (U.S.)
    Frequency of publication: 4 times/month
    Refereed (peer-reviewed)? yes

Contents:
    This is a "scientific general medical journal" that publishes "scientific and clinical articles of interest to a general physician readership."

Submittal specifications:
    Query letter or complete manuscript? manuscript
    Length: variable
    Letter quality required? yes
    Self-addressed, stamped envelope required? no
    Query/manuscript response time: variable
    Time between acceptance and publication: variable
    Photos/illustrations preferred? yes

Terms:
    This periodical does not pay contributors.

13. *The Journal of the Minerals, Metals, and Materials Society*

General information:
    Address: 410 Commonwealth Drive Warrendale, PA 15086
    Managing Editor: James J. Robinson
    Phone: (412)-776-9070
    Fax: (412)-776-3770
    Circulation: 14,000 (approximately)
    Frequency of publication: monthly
    Refereed (peer-reviewed)? no

Contents:
    "JOM presents technical articles, features, overviews, and professional affairs articles on materials science and engineering. The journal is written for and by practitioners in this field (B.S. minimum readership; many are Ph.D.s)."

Submittal specifications:
    Query letter or complete manuscript? query
    Length: 100 to 200 words
    Letter quality required? yes
    Self-addressed, stamped envelope required? no
    Query/manuscript response time: 4 weeks
    Time between acceptance and publication: varies, 2 to 8 months
    Photos/illustrations preferred? yes

Terms:
 Payment: up to $350. However: "JOM is not a viable market for
  freelance writers. Most writing that is paid for is done by
  invitation. We commission and purchase less than 10 articles
  per year. The majority of articles are written gratis. Our
  audience is quite specialized."
 Rights bought: all
 When paid? on publication
 Pay for photos/illustrations: no

14. *Modern Paint and Coatings*

General information:
 Address: 6255 Barfield Road Atlanta, GA 30328
 Editor-in-Chief: Larry Anderson
 Phone: (404)-256-9800
 Fax: (404)-256-3116
 Circulation: 13,000 (approximately)
 Frequency of publication: monthly
 Refereed (peer-reviewed)? no

Contents:
 "Technical articles and case histories about paint manufacturing,
 coatings performance, and regulatory compliance."

Submittal specifications:
 Query letter or complete manuscript? manuscript
 Length: 2000 words
 Letter quality required? yes
 Self-addressed, stamped envelope required? yes
 Query/manuscript response time: varies
 Time between acceptance and publication: 3 months
 Photos/illustrations preferred? yes

Terms:
 "We buy very few articles; most are supplied free by raw
 material suppliers, etc. Freelance fees would be negotiated individually."

15. *The New England Journal of Medicine*

General information:
 Address: 10 Shattuck Street Boston, MA 02115-6094
 Editor-in-Chief: Arnold S. Relman, M.D.
 Phone: (617)-734-9800
 Fax: (617)-734-4457
 Circulation: 230,000
 Refereed (peer-reviewed)? yes

Contents:
>The types of pieces published are: "clinical research articles, medical review articles, brief reports, case records, editorials, correspondence, and opinion pieces."

Submittal specifications:
>Query letter or complete manuscript? manuscript
>Letter quality required? yes
>Self-addressed, stamped envelope required? yes
>Query/manuscript response time: varies
>Time between acceptance and publication: 2 months
>Photos/illustrations preferred? "no preference"

Terms:
>"No payments to authors."

16. *Pit and Quarry*

General information:
>Address: 7500 Old Oak Blvd. Cleveland, OH 44130
>Editor-in-Chief: Bob Drake
>Phone: (216)-891-2746
>Fax: (216)-891-2675
>Circulation: 24,000
>Frequency of publication: monthly
>Refereed (peer-reviewed)? no

Contents:
>"Technical and case-study features dealing with environmental and operational aspects of non-metallic minerals mining."

Submittal specifications:
>Query letter or complete manuscript? manuscript
>Length: 1500 words
>Letter quality required? yes
>Self-addressed, stamped envelope required? yes
>Query/manuscript response time: 3 weeks
>Time between acceptance and publication: 2 to 4 months
>Photos/illustrations preferred? yes

Terms:
>Payment: $50 to $150 per published page, including photos/illustrations
>Kill fee: none
>When paid? on publication
>Pay for photos/illustrations: included in above fees.

17. *Plant/Operations Progress*

General information:
    Address: 345 East 47th Street New York, NY 10017
    Editor-in-Chief: Mark D. Rosenzweig
    Phone: (212)-705-7576
    Fax: (212)-752-3294
    Circulation: 4000
    Frequency of publication: quarterly
    Refereed (peer-reviewed)? no

Contents:
    "Practical articles related to the design and operation of chemical
    process plants."

Submittal specifications:
    Query letter or complete manuscript? manuscript
    Letter quality required? no
    Self-addressed, stamped envelope required? no
    Query/manuscript response time: 2 to 4 weeks
    Time between acceptance and publication: 4 to 6 months
    Photos/illustrations preferred? yes

Terms:
    This periodical does not pay contributors.

18. *Pollution Engineering*

General information:
    Address: 1350 East Touhy Avenue P.O. Box 5080 Des Plaines, IL
        60017-5080
    Editor-in-Chief: Diane Pirocanac
    Phone: (708)-635-8800
    Fax: (708)-390-2752
    Circulation: 50,100
    Frequency of publication: monthly
    Refereed (peer-reviewed)? no
Contents:
    [Periodical's editor did not complete this part of the questionnaire.
    However, from experience, I know that this periodical publishes fea-
    ture articles relating to every aspect of pollution control. The articles
    are usually written by, and geared to, engineers, scientists, and other
    pollution control specialists employed by industries, consulting firms,
    and governmental agencies.]

Submittal specifications:
    Query letter or complete manuscript? either, depending on proposed article
    Letter quality required? no ("$5^1/_4$-in. floppy disk in ASCII preferred")
    Self-addressed, stamped envelope required? no
    Query/manuscript response time: 4 to 6 weeks
    Time between acceptance and publication: 1 to 6 months
    Photos/illustrations preferred? yes

Terms:
    This periodical does not pay contributors.

19. *TAPPI Journal*

General information:
    Address: P.O. Box 105113 Atlanta, GA 30348
    Editor-in-Chief: Matthew Coleman
    Phone: (404)-446-1400
    Fax: (404)-446-6947
    Circulation: 35,000
    Frequency of publication: monthly
    Refereed (peer-reviewed)? yes

Contents:
    Periodical publishes articles on these subjects: papermaking, pulping, engineering, maintenance, environmental, and management.

Submittal specifications:
    Query letter or complete manuscript? manuscript
    Length: 3000 to 5000 words
    Letter quality required? yes
    Self-addressed, stamped envelope required? no
    Query/manuscript response time: 4 weeks
    Time between acceptance and publication: 6 months
    Photos/illustrations preferred? yes

Terms:
    "Negotiable"

Note: I obtained the above information from questionnaires I sent to the editors of selected technical periodicals in early 1991. The data shown are as complete as the responses I received. Though far from all-inclusive, the periodicals listed nonetheless represent a cross-section of today's market. By the time you read this, the data may have changed. To be sure that your query is properly written and directed, you should contact the periodical before submitting to it.

6   The Business of Technical Writing

No man but a blockhead ever wrote except for money.
Samuel Johnson,

Boswell's Life of Dr. Johnson, Volume II

If you write technical articles or books, or plan to, you need to be familiar with the legal and business aspects of writing. This applies regardless of whether writing is your vocation or your avocation. If you're a sometime author, you might argue that these matters needn't concern you.

"Shouldn't publishers and periodicals be the ones to deal with copyright and other legal matters?" you might ask. "Anyway, should I care about the business end — expenses, income taxes, and so forth? I'm not a professional writer."

Nevertheless, the truth is that you do need to care about these things. You can't pass the buck to the magazine or publisher. For example, if you've written a journal article that includes a lot of previously published material for which you haven't given proper credit, you could be inviting a copyright infringement suit. Even if the publisher or periodical should be named as a codefendant in such a suit, you could still be liable for part of the damages if you lost. And even if you won the suit, you'd probably still have to pay a share of legal fees. As discussed in Chapter 4, publishers try to discourage these suits by including in their standard book contracts a clause that makes the author responsible for assuring that no part of the book violates *any* laws — copyright, libel, or otherwise. In other words, your "amateur status" as a writer affords you no protection whatsoever.

The business of writing is also worthy of your attention. Unless you earn absolutely nothing from your writing, you will have to report your writing income to the Internal Revenue Service, as well as to your state and local revenue authorities; no amount is too small to report. Don't for a moment think that you can "forget" to pay the tax collectors. These days, periodicals and publishers routinely obtain each author's Social Security number and report it to the taxing authorities, along with any payments made to that author. Any writer who neglects to report these payments (however modest they may be) on his tax return is inviting an audit.

Finally, an author should be aware of the deductions she is entitled to take by virture of earning writing income. These include expenses for such things as maintaining an office, using a vehicle, and traveling to technical conferences.

In this chapter, we'll discuss these and other legal and business issues. By the time you're finished reading it, you will have obtained an overview of these matters — matters that professional writers must deal with day to day. However, an overview is all this chapter provides. For more detailed information, you should consult either the references listed at the end of this chapter or, better yet, a tax accountant and/or an attorney who specializes in publishing law.

TAXES AND THE WRITER

As discussed earlier, every writer who earns money from writing will eventually have to report that income to the tax collectors. At the very least, this would include the Internal Revenue Service. But for many (if not most) of us, the income also would have to be reported to state and local authorities. No amount, however small, is exempt.*

This doesn't mean that you would have to spend a great deal of effort documenting writing income on your tax return. If the amount is relatively small (say, less than $1000), you can just report it on your return as "other income" (line 22 on 1990 Form 1040). The tax collector will be satisfied, and you'll be able to sleep better at night. But, if you merely report writing income and fail to list writing *deductions*, you'll likely be depriving yourself of hard-earned dollars. And if you look closely, you'll find several such deductions to offset your writing income.

---

* Because many state and some local income tax codes and forms are patterned after that of the Internal Revenue Service, only the IRS will be mentioned in the rest of the income tax discussion. This is done to make things simpler. However, the deductions and other subjects covered may not necessarily apply to state and local income taxes.

Writing Deductions Big and Small

"But wait a minute," you say. "How can I take writing business-type deductions if I'm not a full-time writer?" According to the IRS, you don't have to depend on writing for your entire living to qualify as a part-time writer. If you can demonstrate to the tax collectors that your writing activities constitute a business, then you may complete "Schedule C. Profit or Loss from Business" and attach it to your tax return (see Figure 6.1). How do you demonstrate this? The IRS has devised seven tests by which they can measure your "material participation" in the activity. Most of these tests mandate a certain minimum level of participation on your part in the activity (e.g., 500 hours during the tax year). If you meet any one of these tests, you would be a "material participant" in the activity and, accordingly, the IRS would consider the activity to be a business for tax purposes. However, if you cannot meet any of the seven tests, your business would be a "passive activity". Generally, this means that you can deduct losses from the business only to the extent of income earned therefrom.[1]

Supposing you do qualify, what expenses may you deduct? There are a large number, ranging from paper to printers. They include:

- Office supplies
- Periodical subscriptions
- Books
- Legal and professional services
- Professional association dues
- Travel expenses
- Telephone, telefaxing, and photocopying expenses
- Vehicle expenses
- Office equipment
- "Home office expense"

Let's take a closer look at each of these.

Office Supplies

This includes such items as paper, writing implements, ribbons, envelopes, and small equipment items (staplers, tape dispensers) that have useful lives of less than 1 year.

Periodical Subscriptions

Covered here would be any periodicals that relate to the writing business (e.g., *Writer's Digest*) or to subjects that you would regularly write about. The latter could include technical journals related to your field or to any other specialty, as long as they are connected, directly or indirectly, to your writing.

SCHEDULE C (Form 1040) Department of the Treasury Internal Revenue Service (O)	**Profit or Loss From Business** (Sole Proprietorship) Partnerships, Joint Ventures, Etc., Must File Form 1065. ▶ Attach to Form 1040 or Form 1041. ▶ See Instructions for Schedule C (Form 1040).	OMB No. 1545-0074 19**90** Attachment Sequence No. **09**

Name of proprietor	Social security number (SSN)

A Principal business or profession, including product or service (see Instructions)	**B** Enter principal business code (from page 2) ▶

C Business name and address ▶ ....
(include suite or room no.)

D Employer ID number (Not SSN)

E Accounting method: **(1)** ☐ Cash **(2)** ☐ Accrual **(3)** ☐ Other (specify) ▶ ....

F Method(s) used to value closing inventory: **(1)** ☐ Cost **(2)** ☐ Lower of cost or market **(3)** ☐ Other (attach explanation) **(4)** ☐ Does not apply (if checked, go to line H) | Yes | No

G Was there any change in determining quantities, costs, or valuations between opening and closing inventory? (If "Yes," attach explanation.)

H Are you deducting expenses for business use of your home? (If "Yes," see Instructions.)

I Did you "materially participate" in the operation of this business during 1990? (If "No," see Instructions for limitations on losses.)

J If this is the first Schedule C filed for this business, check here . . . . ▶ ☐

Part I Income

1	Gross receipts or sales. *Caution: If this income was reported to you on Form W-2 and the "Statutory employee" box on that form was checked, see the Instructions and check here* . . . . ▶ ☐	1
2	Returns and allowances	2
3	Subtract line 2 from line 1. Enter the result here	3
4	Cost of goods sold (from line 38 on page 2)	4
5	Subtract line 4 from line 3 and enter the **gross profit** here	5
6	Other income, including Federal and state gasoline or fuel tax credit or refund (see Instructions)	6
7	Add lines 5 and 6. This is your **gross income** . . . . ▶	7

Part II Expenses

8	Advertising	8	21 Repairs and maintenance . . .	21
9	Bad debts from sales or services (see Instructions)	9	22 Supplies (not included in Part III) .	22
10	Car and truck expenses (attach **Form 4562**)	10	23 Taxes and licenses	23
11	Commissions and fees . . .	11	24 Travel, meals, and entertainment:	
12	Depletion	12	**a** Travel	24a
13	Depreciation and section 179 expense deduction (not included in Part III) (see Instructions) . .	13	**b** Meals and entertainment .	
14	Employee benefit programs (other than on line 19)	14	**c** Enter 20% of line 24b subject to limitations (see Instructions) . .	
15	Insurance (other than health) . .	15	**d** Subtract line 24c from line 24b .	24d
16	Interest:		25 Utilities	25
	a Mortgage (paid to banks, etc.) .	16a	26 Wages (less jobs credit)	26
	b Other	16b	27a Other expenses (list type and amount):	
17	Legal and professional services .	17	....	
18	Office expense	18	....	
19	Pension and profit-sharing plans .	19	....	
20	Rent or lease (see Instructions):		....	
	a Vehicles, machinery, and equip. .	20a	....	
	b Other business property . . .	20b	27b Total other expenses	27b

28	Add amounts in columns for lines 8 through 27b. These are your **total expenses** . . . . ▶	28
29	**Net profit or (loss)**. Subtract line 28 from line 7. If a profit, enter here and on Form 1040, line 12. Also enter the net profit on Schedule SE, line 2 (statutory employees, see Instructions). If a loss, you MUST go on to line 30 (fiduciaries, see Instructions)	29
30	If you have a loss, you MUST check the box that describes your investment in this activity (see Instructions) . .	30a ☐ All investment is at risk. 30b ☐ Some investment is not at risk.
	If you checked 30a, enter the loss on Form 1040, line 12, and Schedule SE, line 2 (statutory employees, see Instructions). If you checked 30b, you MUST attach **Form 6198**.	

For Paperwork Reduction Act Notice, see Form 1040 Instructions. Schedule C (Form 1040) 1990

Figure 6.1. "Schedule C. Profit or Loss From Business" form.

Books

Deductible books include references (both the general, like dictionaries, and the specialized, such as handbooks) and others related to your writing. However, if a book has a useful life of more than 1 year (a set of encyclopedias, for instance) it would have to be either (1) "depreciated" over the IRS-allowed lifetime or (2) treated as a "Section 179

Schedule C (Form 1040) 1990 | | Page **2**

Part III Cost of Goods Sold (See Instructions.)

31 Inventory at beginning of year. (If different from last year's closing inventory, attach explanation.)	31	
32 Purchases less cost of items withdrawn for personal use	32	
33 Cost of labor. (Do not include salary paid to yourself.)	33	
34 Materials and supplies	34	
35 Other costs	35	
36 Add lines 31 through 35	36	
37 Inventory at end of year.	37	
38 **Cost of goods sold.** Subtract line 37 from line 36. Enter the result here and on page 1, line 4	38	

Part IV Principal Business or Professional Activity Codes

Locate the major category that best describes your activity. Within the major category, select the activity code that most closely identifies the business or profession that is the principal source of your sales or receipts. **Enter this 4-digit code on page 1, line B.** For example, a grocery store is under the major category of "Retail Trade," and the code is "3210." (**Note:** If your principal source of income is from farming activities, you should file **Schedule F** (Form 1040), Farm Income and Expenses.)

Construction

Code

0018 Operative builders (for own account)

General contractors

0034 Residential building
0059 Nonresidential building
0075 Highway and street construction
3889 Other heavy construction (pipe laying, bridge construction, etc.)

Building trade contractors, including repairs

0232 Plumbing, heating, air conditioning
0257 Painting and paper hanging
0273 Electrical work
0299 Masonry, dry wall, stone, tile
0414 Carpentering and flooring
0430 Roofing, siding, and sheet metal
0455 Concrete work
0885 Other building trade contractors (excavation, glazing, etc.)

Manufacturing, Including Printing and Publishing

0638 Food products and beverages
0653 Textile mill products
0679 Apparel and other textile products
0695 Leather, footware, handbags, etc.
0810 Furniture and fixtures
0836 Lumber and other wood products
0851 Printing and publishing
0877 Paper and allied products
1032 Stone, clay, and glass products
1057 Primary metal industries
1073 Fabricated metal products
1099 Machinery and machine shops
1115 Electric and electronic equipment
1883 Other manufacturing industries

Mining and Mineral Extraction

1511 Metal mining
1537 Coal mining
1552 Oil and gas
1719 Quarrying and nonmetallic mining

Agricultural Services, Forestry, Fishing

1933 Crop services
1958 Veterinary services, including pets
1974 Livestock breeding
1990 Other animal services
2113 Farm labor and management services
2212 Horticulture and landscaping
2238 Forestry, except logging
0836 Logging
2246 Commercial fishing
2469 Hunting and trapping

Wholesale Trade—Selling Goods to Other Businesses, Etc.

Durable goods, including machinery, equipment, wood, metals, etc.

2618 Selling for your own account
2634 Agent or broker for other firms—more than 50% of gross sales on commission

Nondurable goods, including food, fiber, chemicals, etc.

2659 Selling for your own account

2675 Agent or broker for other firms—more than 50% of gross sales on commission

Retail Trade—Selling Goods to Individuals and Households

3012 Selling door-to-door, by telephone or party plan, or from mobile unit
3038 Catalog or mail order
3053 Vending machine selling

Selling From Showroom, Store, or Other Fixed Location

Food, beverages, and drugs

3079 Eating places (meals or snacks)
3086 Catering services
3095 Drinking places (alcoholic beverages)
3210 Grocery stores (general line)
0612 Bakeries selling at retail
3236 Other food stores (meat, produce, candy, etc.)
3251 Liquor stores
3277 Drug stores

Automotive and service stations

3319 New car dealers (franchised)
3335 Used car dealers
3517 Other automotive dealers (motorcycles, recreational vehicles, etc.)
3533 Tires, accessories, and parts
3558 Gasoline service stations

General merchandise, apparel, and furniture

3715 Variety stores
3731 Other general merchandise stores
3756 Shoe stores
3772 Men's and boys' clothing stores
3913 Women's ready-to-wear stores
3921 Women's accessory and specialty stores and furriers
3939 Family clothing stores
3954 Other apparel and accessory stores
3970 Furniture stores
3996 TV, audio, and electronics
3988 Computer and software stores
4119 Household appliance stores
4317 Other home furnishing stores (china, floor coverings, etc.)
4333 Music and record stores

Building, hardware, and garden supply

4416 Building materials dealers
4432 Paint, glass, and wallpaper stores
4457 Hardware stores
4473 Nurseries and garden supply stores

Other retail stores

4614 Used merchandise and antique stores (except motor vehicle parts)
4630 Gift, novelty, and souvenir shops
4655 Florists
4671 Jewelry stores
4697 Sporting goods and bicycle shops
4812 Boat dealers
4838 Hobby, toy, and game shops
4853 Camera and photo supply stores
4879 Optical goods stores
4895 Luggage and leather goods stores
5017 Book stores, excluding newsstands
5033 Stationery stores
5058 Fabric and needlework stores
5074 Mobile home dealers
5090 Fuel dealers (except gasoline)
5884 Other retail stores

Finance, Insurance, Real Estate, and Related Services

5520 Real estate agents or brokers
5579 Real estate property managers
5710 Subdividers and developers, except cemeteries
5538 Operators and lessors of buildings, including residential
5553 Operators and lessors of other real property
5702 Insurance agents or brokers
5744 Other insurance services
6064 Security brokers and dealers
6080 Commodity contracts brokers and dealers, and security and commodity exchanges
6130 Investment advisors and services
6148 Credit institutions and mortgage bankers
6155 Title abstract offices
5777 Other finance and real estate

Transportation, Communications, Public Utilities, and Related Services

6114 Taxicabs
6312 Bus and limousine transportation
6361 Other highway passenger transportation
6338 Trucking (except trash collection)
6395 Courier or package delivery services
6510 Trash collection without own dump
6536 Public warehousing
6551 Water transportation
6619 Air transportation
6635 Travel agents and tour operators
6650 Other transportation services
6676 Communication services
6692 Utilities, including dumps, snowplowing, road cleaning, etc.

Services (Personal, Professional, and Business Services)

Hotels and other lodging places

7096 Hotels, motels, and tourist homes
7211 Rooming and boarding houses
7237 Camps and camping parks

Laundry and cleaning services

7419 Coin-operated laundries and dry cleaning
7435 Other laundry, dry cleaning, and garment services
7450 Carpet and upholstery cleaning
7476 Janitorial and related services (building, house, and window cleaning)

Business and/or personal services

7617 Legal services (or lawyer)
7633 Income tax preparation
7658 Accounting and bookkeeping
7518 Engineering services
7682 Architectural services
7708 Surveying services
7245 Management services
7260 Public relations
7286 Consulting services
7716 Advertising, except direct mail
7732 Employment agencies and personnel supply
7799 Consumer credit reporting and collection services

7856 Mailing, reproduction, commercial art and photography, and stenographic services
7872 Computer programming, processing, data preparation, and related services
7922 Computer repair, maintenance, and leasing
7773 Equipment rental and leasing (except computer or automotive)
7914 Investigative and protective services
7880 Other business services

Personal services

8110 Beauty shops (or beautician)
8318 Barber shop (or barber)
8334 Photographic portrait studios
8532 Funeral services and crematories
8714 Child day care
8730 Teaching or tutoring
8755 Counseling (except health practitioners)
8771 Ministers and chaplains
6882 Other personal services

Automotive services

8813 Automotive rental or leasing, without driver
8839 Parking, except valet
8953 Automotive repairs, general and specialized
8896 Other automotive services (wash, towing, etc.)

Miscellaneous repair, except computers

9019 TV and audio equipment repair
9035 Other electrical equipment repair
9050 Reupholstery and furniture repair
2881 Other equipment repair

Medical and health services

9217 Offices and clinics of medical doctors (MDs)
9233 Offices and clinics of dentists
9258 Osteopathic physicians and surgeons
9241 Podiatrists
9274 Chiropractors
9290 Optometrists
9415 Registered and practical nurses
9431 Other health practitioners
9456 Medical and dental laboratories
9472 Nursing and personal care facilities
9886 Other health services

Amusement and recreational services

8557 Physical fitness facilities
9597 Motion picture and video production
9688 Motion picture and tape distribution and allied services
9613 Videotape rental
9639 Motion picture theaters
9670 Bowling centers
9696 Professional sports and racing, including promoters and managers
9811 Theatrical performers, musicians, agents, producers, and related services
9837 Other amusement and recreational services

8888 Unable to classify

Figure 6.1 continued.

deduction". (We'll discuss both depreciation and Section 179 deductions below.) Examples of books in the one-year-or-less class are the *Writer's Market,* the *Literary Market Place,* and the *Physician's Desk Reference.* Since each of these references is updated annually, one can legitimately treat them as expenses each year on one's tax return.

Legal and Professional Services

Many (if not most) professional writers have agents to represent them in their dealings with publishers and periodicals. The more successful wordsmiths also may have attorneys, business managers, tax consultants, and others on retainer. Fees paid to these individuals would be deductible. Even if you're not a professional writer, you may have needed to consult, say, an attorney regarding some contract matter during the year. If so, his or her fee would be deductible. So would any fees you had paid to a tax preparer. (Tax preparation fees are deductible, regardless of whether you file a Schedule C form. But the deductible portion of these fees and your other "job expenses and miscellaneous deductions" would only be that amount *in excess* of 2% of your adjusted gross income.)

Professional Association Dues

This category includes membership dues and fees paid to any organization related to writing (e.g., the Author's Guild), to your specialty, or to any other subject connected to your writing. For instance, if you are an attorney not in private practice, you could deduct your bar association dues on Schedule C if you wrote in your spare time about matters related to the law. Otherwise, you wouldn't be able to deduct these dues unless your other miscellaneous deductions already exceeded 2% of your adjusted gross income.

Travel Expenses

Any expenses — airline tickets, lodging, public transportation — connected with writing-related trips would be deductible. Suppose that your publisher has asked you to attend a trade fair at which your new book will be promoted and sold. Suppose further that the publisher is having "cash flow problems" and cannot pay your travel expenses to said fair. Unfortunately, you'd have to foot the bill yourself. Fortunately, these expenses would be an allowable deduction. What if you had planned to take an extra day or two to do some sightseeing during this trip? In that case, you wouldn't be allowed to deduct the total cost of the trip, but just that portion associated with the trade fair. But there is a catch or two. For instance, IRS allows a taxpayer to deduct only 80% of the cost of business meals and entertainment. Furthermore, IRS rules don't allow the deduction of business travel expenses to locations outside the U.S., Canada, Mexico, and Jamaica, unless you can show that it is reasonable for you to conduct your business there. There are other exceptions to business deductions as well. For more information, consult Form 1040 or Publication 334 ("Tax Guide for Small Businesses", Reference 2).

Vehicle Expenses

Whenever you use your own vehicle to take writing-related trips (local or long-distance), you can deduct the cost of using it. The simplest method for figuring this deduction (and probably the best one for part-time writers) is to do it on a mileage basis. The IRS now allows you to take a deduction of $0.26 per mile — an allowance that has increased steadily over the years.[1] Of course, you would have to record your mileage carefully, indicating the date, destination, odometer readings, and purpose for each trip. What kind of trips can you include? Visits to the library, bookstores, and other locations for research purposes are permitted. So are meetings with your lawyer, agent, and other advisors. Trips to stores to pick up supplies are also kosher.... You get the idea.

Telephone, Telefaxing, Photocopying, etc.

This deduction embraces all expenses associated with *communication*. Whenever you phone your publisher, fax him a letter, or photocopy a manuscript before sending it to him, you may deduct the cost of doing so. If you use a phone credit card or obtain faxing and photocopying services from an outside source, you can easily document these expenses. However, if you use your home phone and/or fax machine to make business calls, you would need to figure these deductions differently. See Form 1040 or Publication 334 for more details.

Office Equipment

Unlike office supplies, which cover only small items that are used up or wear out in a year or less, the "office equipment" category covers larger items, such as furniture and such devices as typewriters, photocopiers, paper cutters and shredders, postage scales and meters, telephones, and computer systems. If you plan to take this deduction, you will have to complete Form 4562, "Depreciation and Amortization (Including Information on Listed Property)" and attach it to your tax return (see Figure 2).

Note that there are two ways to account for office equipment on this form. First, you can figure a depreciation allowance, via the General Depreciation System (GDS or MACRS), the Alternative Depreciation System (ADS or alternate MACRS), the Accelerated Cost Recovery System (ACRS, for certain assets placed in service before 1990), or Section 168(f). Both the GDS and ADS deductions are based on the IRS-allowed property life (which ranges from 3 to 40 years) and the depreciation method used (i.e., straight line, declining balance, etc.). According to the IRS, office machinery (computers, etc.) are considered 3-year property, while office furniture and fixtures are included in the 5-year property category.[2]

In lieu of taking a depreciation allowance, you may deduct all or part of the cost of an asset as an *expense* on your return. Known as a "Section 179 deduction," this deduction cannot be taken until the asset is placed in service. Note that the Section 179 deduction is limited to $10,000 per tax year *per taxpayer,* not per business. However, "for each dollar of cost of Section 179 property placed in service in excess of $200,000 in a tax year, the $10,000 maximum is reduced...by one dollar."[2] Nonetheless, this deduction is very handy for offsetting one's taxable income, especially in years when it's higher than average.

However, there are other restrictions on taking the Section 179 deduction. The IRS has designated certain assets as "listed property". These include "automobiles, certain other vehicles, cellular telephones, computers, and property used for entertainment, recreation, or amusement." Of these, writers would be most interested in cellular telephones and computers. For these assets, one must attest — and be able to prove — that they were used "more than 50% in a trade or business;" and only that portion that was used in a trade or business would be eligible for the Section 179 deduction. In these cases, the deduction would be figured on a pro rata basis (see Figure 6.2).

"Home Office Expense"

The home office expense (H.O.E.) deduction (not to be confused with the "office supplies" or "office equipment" deductions) is one that the IRS examines very closely these days. This increased scrutiny was brought on by the widespread abuse of this deduction during recent years. Before the passage of the 1986 Tax Reform Act, nearly anyone, self-employed or not, could take a home office deduction, even someone who had an office at his place of business. A taxpayer might claim this expense because he regularly had to bring work home on evenings and weekends. Never mind that "regularly" might mean once in a blue moon. To close this loophole, the IRS tightened the requirements for allowing a home office expense.

The most recent IRS guidance on this deduction is contained in their Publication 587, "Business Use of Your Home."[3] As explained therein, an H.O.E. deduction is allowed *only* if a part of your home is used "exclusively and regularly" for business purposes. "Exclusively" means just that. For example, if you use a den for an office 90% of the time, but use it the rest of the time for personal pursuits, you can't take the H.O.E. deduction. In addition, if you are an employee, the business use of your home must be "for the convenience of your employer." Also, the office must be used "regularly"; that is, on a "continuing basis". Although the IRS doesn't define what "regular" means, it certainly doesn't mean once a month or twice a quarter.

Despite these restrictions, the H.O.E. can be an attractive deduction, as you may be allowed to deduct a portion of the total cost of owning

Form **4562**	**Depreciation and Amortization**	OMB No. 1545-0172
Department of the Treasury Internal Revenue Service	**(Including Information on Listed Property)** ▶ See separate instructions. ▶ Attach this form to your return.	**1990** Attachment Sequence No. **67**

Name(s) shown on return	Identifying number

Business or activity to which this form relates

Part I Election To Expense Certain Tangible Property (Section 179) *(Note: If you have any "Listed Property," also complete Part V.)*

1 Maximum dollar limitation (see instructions)	**1**	$10,000
2 Total cost of section 179 property placed in service during the tax year (see instructions)	**2**	
3 Threshold cost of section 179 property before reduction in limitation	**3**	$200,000
4 Reduction in limitation—Subtract line 3 from line 2, but do not enter less than -0-	**4**	
5 Dollar limitation for tax year—Subtract line 4 from line 1, but do not enter less than -0-	**5**	

(a) Description of property	(b) Cost	(c) Elected cost	
6			

7 Listed property—Enter amount from line 26	**7**		
8 Total elected cost of section 179 property—Add amounts in column (c), lines 6 and 7		**8**	
9 Tentative deduction—Enter the lesser of line 5 or line 8		**9**	
10 Carryover of disallowed deduction from 1989 (see instructions)		**10**	
11 Taxable income limitation—Enter the lesser of taxable income or line 5 (see instructions)		**11**	
12 Section 179 expense deduction—Add lines 9 and 10, but do not enter more than line 11		**12**	
13 Carryover of disallowed deduction to 1991—Add lines 9 and 10, less line 12 ▶	**13**		

Note: *Do not use Part II or Part III below for automobiles, certain other vehicles, cellular telephones, computers, or property used for entertainment, recreation, or amusement (listed property). Instead, use Part V for listed property.*

Part II MACRS Depreciation For Assets Placed in Service ONLY During Your 1990 Tax Year (Do Not Include Listed Property)

(a) Classification of property	(b) Mo. and yr. placed in service	(c) Basis for depreciation (Business use only—see instructions)	(d) Recovery period	(e) Convention	(f) Method	(g) Depreciation deduction
14 General Depreciation System (GDS) (see instructions):						
a 3-year property						
b 5-year property						
c 7-year property						
d 10-year property						
e 15-year property						
f 20-year property						
g Residential rental property			27.5 yrs.	MM	S/L	
			27.5 yrs.	MM	S/L	
h Nonresidential real property			31.5 yrs.	MM	S/L'	
			31.5 yrs.	MM	S/L	
15 Alternative Depreciation System (ADS) (see instructions):						
a Class life					S/L	
b 12-year			12 yrs.		S/L	
c 40-year			40 yrs.	MM	S/L	

Part III Other Depreciation (Do Not Include Listed Property)

16 GDS and ADS deductions for assets placed in service in tax years beginning before 1990 (see instructions).	**16**	
17 Property subject to section 168(f)(1) election (see instructions)	**17**	
18 ACRS and other depreciation (see instructions)	**18**	

Part IV Summary

19 Listed property—Enter amount from line 25	**19**	
20 Total—Add deductions on line 12, lines 14 and 15 in column (g), and lines 16 through 19. Enter here and on the appropriate lines of your return. (Partnerships and S corporations—see instructions) . .	**20**	
21 For assets shown above and placed in service during the current year, enter the portion of the basis attributable to section 263A costs (see instructions).	**21**	

For Paperwork Reduction Act Notice, see page 1 of the separate instructions.   Form **4562** (1990)

Figure 6.2. "Depreciation and Amortization (Including Information on Listed Property)" form.

your home as well as all direct costs associated with the office. How is this deduction figured? First, it consists of two parts. The first part is comprised of those costs directly associated with improving and maintaining the office area itself. This might include painting, redoing the floors, fixing broken windows, and the like. The other part is the "indirect" H.O.E. cost — that portion of your total home ownership cost

Form 4562 (1990)   Page **2**

Part V   **Listed Property.—Automobiles, Certain Other Vehicles, Cellular Telephones, Computers, and Property Used for Entertainment, Recreation, or Amusement**

If you are using the standard mileage rate or deducting vehicle lease expense, complete columns (a) through (c) of Section A, all of Section B, and Section C if applicable.

Section A.—Depreciation   (**Caution:** *See instructions for limitations for automobiles.*)

22a Do you have evidence to support the business use claimed? ☐ **Yes** ☐ **No** | **22b** If "Yes," is the evidence written? ☐ **Yes** ☐ **No**

(a) Type of property (list vehicles first)	(b) Date placed in service	(c) Business use percentage	(d) Cost or other basis	(e) Basis for depreciation (business use only)	(f) Recovery period	(g) Method/ Convention	(h) Depreciation deduction	(l) Elected section 179 cost
23 *Property used more than 50% in a trade or business:*								
		%						
		%						
		%						
24 *Property used 50% or less in a trade or business:*								
		%				S/L –		
		%				S/L –		
		%				S/L –		

25 Add amounts in column (h). Enter the total here and on line 19, page 1 . . . . **25**

26 Add amounts in column (i). Enter the total here and on line 7, page 1 . . . . **26**

Section B.—Information Regarding Use of Vehicles—*If you deduct expenses for vehicles:*

● *Always complete this section for vehicles used by a sole proprietor, partner, or other "more than 5% owner," or related person.*

● *If you provided vehicles to your employees, first answer the questions in Section C to see if you meet an exception to completing this section for those vehicles.*

	(a) Vehicle 1		(b) Vehicle 2		(c) Vehicle 3		(d) Vehicle 4		(e) Vehicle 5		(f) Vehicle 6	
27 Total business miles driven during the year (DO NOT include commuting miles) . . .												
28 Total commuting miles driven during the year												
29 Total other personal (noncommuting) miles driven												
30 Total miles driven during the year—Add lines 27 through 29												
	Yes	No	Yes	No	Yes	No	Yes	No	Yes	No	Yes	No
31 Was the vehicle available for personal use during off-duty hours?												
32 Was the vehicle used primarily by a more than 5% owner or related person? . . .												
33 Is another vehicle available for personal use?												

Section C.—Questions for Employers Who Provide Vehicles for Use by Their Employees

*(Answer these questions to determine if you meet an exception to completing Section B. **Note:** Section B must always be completed for vehicles used by sole proprietors, partners, or other more than 5% owners or related persons.)*

	Yes	No
34 Do you maintain a written policy statement that prohibits all personal use of vehicles, including commuting, by your employees?		
35 Do you maintain a written policy statement that prohibits personal use of vehicles, except commuting, by your employees? (See instructions for vehicles used by corporate officers, directors, or 1% or more owners.)		
36 Do you treat all use of vehicles by employees as personal use?		
37 Do you provide more than five vehicles to your employees and retain the information received from your employees concerning the use of the vehicles?.		
38 Do you meet the requirements concerning qualified automobile demonstration use (see instructions)?		

Note: *If your answer to 34, 35, 36, 37, or 38 is "Yes," you need not complete Section B for the covered vehicles.*

Part VI   **Amortization**

(a) Description of costs	(b) Date amortization begins	(c) Amortizable amount	(d) Code section	(e) Amortization period or percentage	(f) Amortization for this year
39 Amortization of costs that begins during your 1990 tax year:					
40 Amortization of costs that began before 1990.				**40**	
41 Total. Enter here and on "Other Deductions" or "Other Expenses" line of your return				**41**	

Figure 6.2 continued.

one can attribute to the home office area. This cost is figured by multiplying the total home ownership cost by the ratio of the home office area to the total area of the home.

> **Example:** a home office occupies 150 sq. ft. in a 1500-sq. ft. home. The total home ownership cost for the tax year was $12,000. Hence, the indirect H.O.E. cost would be: 150/1500 ¥ $12,000 = $1200. However, it cost

another $650 to paint the office walls and refinish the floor. The total H.O.E. deduction would therefore be: $1200 + $650 = $1850.

What kinds of expenditures does the "total home ownership cost" include? The IRS suggests several: real estate taxes, deductible mortgage interest, casualty losses, rent, utilities and services, insurance, repairs, security system, and depreciation. Reference 3 provides more information about these expenditures and what they include.

Recordkeeping Requirements

The IRS requires all taxpayers to keep good records. This is especially true of the self-employed and others who are in business, full or part-time. These records must not only address tax-deductible expenditures, but all other expenditures and income as well. Although the IRS doesn't mandate a particular type of recordkeeping, they do suggest that the taxpayer keep canceled checks, receipts, and other evidence of expenses paid. Those in business should also keep records of receipts — amounts, dates, and names of payors. Further, the IRS requires that these records be kept "as long as they are important for any Internal Revenue law." In most cases, this is 3 years from the date the return was due or 2 years from the date the tax was paid, whichever is later. To keep your records organized, a ledger is handy. It needn't be anything fancy, just a book to record income and expenditures. Nor do you need to be a bookkeeper to use it. Simply record, for each entry: the date the amount is paid/received; a brief description of the expenditure/income; the check or other account number; and the amount.

To keep your records up to date, make an entry as soon as possible after receiving/paying an amount. Consider the following (hypothetical) illustration.

Date	Description	Check #	Receipt	Expense
1/30/91	Fr: Acme Publishing For: Book royalties (7/90–12/90)	—	$947.00	
2/5/91	To: Jewel Office Supply For: Copier cartridge	892		$76.88
2/8/91	To: *Cryogenic Journal* For: 1-yr. subscription	902		25.00
2/15/91	Fr: *Law Times* For: Honorarium — expert witness article	—	55.00	

For each expenditure entry in the journal, you should keep a canceled check or other receipt for documentation. Each receipt should list the date, payee, and amount and describe the items purchased. For each income entry, you should retain a statement from the payor that contains the same information. For convenience — and to minimize misplacements — keep the receipts and income statements in separate envelopes. If you're audited (heaven forbid!), at least you'll have your records in order. Finally, the cost of such a ledger is a tax-deductible expense.

Other Taxes to Worry About

Unfortunately, income taxes are not the only taxes that the full- or part-time writer has to be concerned with. The "significant others" include the Social Security tax, occupational privilege (and similar) taxes, and other levies that our duly elected representatives have seen fit to foist upon us. In addition, for those writers who employ others (secretaries, researchers, etc.), there is Workman's Compensation, withholding, and the like to contend with. However, as most of those who write technical material are part-timers, employee-related taxes are usually not a concern.

Social Security Tax

If the net profit from your writing endeavors (line 29, Schedule C) is $400 or more, you will have to complete Schedule SE ("Social Security Self-Employment Tax") and attach it to your return (see Figure 6.3). If the amount of your other wages and self-employment earnings exceeds the maximum amount ($51,300 in 1990), you will not owe any additional Social Security tax; but if it doesn't, you'll have to pay more. How much more will depend both on your net profit and the method you use to compute your tax; that is, "Short Schedule SE" or "Long Schedule SE". If you use the Short Schedule, the tax rate is 15.3% of the amount on line 8 of Schedule SE. However, as Figure 6.3 shows, this amount is somewhat less than your net profit, so that the *effective* tax rate will be something less than 15.3%.

Occupational Privilege Tax/Business License

This tax, levied locally, varies accordingly. Most communities levy a tax on businesses, in the form of licenses or permits. Most of them define "business" according to the commonly understood meaning of the term (i.e., a place where customers or clients come to purchase goods or services). However, some communities expand that definition to include *any* business, even one conducted in a home. The key point is how that definition reads. If a business must involve the comings and goings of customers or clients, then most writers would be exempt, as they generally do their "business" alone; but if the local ordinance includes all businesses, with no exceptions, a writer may very well be liable for paying this tax.

SCHEDULE SE (Form 1040) Department of the Treasury Internal Revenue Service (O)	**Social Security Self-Employment Tax** ▶ See Instructions for Schedule SE (Form 1040). ▶ Attach to Form 1040.	OMB No. 1545-0074 19**90** Attachment Sequence No. 17

Name of person with **self-employment** income (as shown on Form 1040)	Social security number of person with **self-employment** income ▶	

Who Must File Schedule SE

You must file Schedule SE if:

- Your net earnings from self-employment were $400 or more; **OR**
- You were an employee of an electing church or church-controlled organization that paid you wages (church employee income) of $100 or more;

  AND

- Your wages (subject to social security or railroad retirement tax) were less than $51,300.

Exception: If your only self-employment income was from earnings as a minister, member of a religious order, or Christian Science practitioner, AND you filed **Form 4361** and received IRS approval not to be taxed on those earnings, DO NOT file Schedule SE. Instead, write "Exempt–Form 4361" on Form 1040, line 48.

For more information about Schedule SE, see the Instructions.

Note: *Most people can use the short Schedule SE on this page. But, you may have to use the longer Schedule SE on the back.*

Who MUST Use the Long Schedule SE (Section B)

You must use Section B if ANY of the following apply:

- You elect the "optional method" to figure your self-employment tax (see Section B, Part II, and the Instructions);
- You are a minister, member of a religious order, or Christian Science practitioner and you received IRS approval (from **Form 4361**) not to be taxed on your earnings from these sources, but you owe self-employment tax on other earnings;
- You had church employee income of $100 or more that was reported to you on Form W-2;
- You had tip income that is subject to social security tax, but you did not report those tips to your employer; OR
- You were a government employee with wages subject ONLY to the 1.45% Medicare part of the social security tax (Medicare qualified government wages) AND the total of **all** of your wages (subject to social security, railroad retirement, or the 1.45% Medicare tax) plus **all** your earnings subject to self-employment tax is **more** than $51,300.

Section A—Short Schedule SE (Read above to see if you must use the long Schedule SE on the back (Section B).)

1	Net farm profit or (loss) from Schedule F (Form 1040), line 36, and farm partnerships, Schedule K-1 (Form 1065), line 15a	**1**	
2	Net profit or (loss) from Schedule C (Form 1040), line 29, and Schedule K-1 (Form 1065), line 15a (other than farming). See Instructions for other income to report.	**2**	
3	Combine lines 1 and 2. Enter the result	**3**	
4	Multiply line 3 by .9235. If the result is less than $400, **do not** file this schedule; you **do not** owe self-employment tax. . . . . ▶	**4**	
5	Maximum amount of combined wages and self-employment earnings subject to social security or railroad retirement (tier 1) tax for 1990	**5**	$51,300 | 00
6	Total social security wages and tips (from Form(s) W-2) and railroad retirement compensation (tier 1). **Do not** include Medicare qualified government wages on this line	**6**	
7	Subtract line 6 from line 5. Enter the result. If the result is zero or less, **do not** file this schedule; you **do not** owe self-employment tax . . . . ▶	**7**	
8	Enter the **smaller** of line 4 or line 7	**8**	
9	Rate of tax	**9**	×.153
10	**Self-employment tax.** If line 8 is $51,300, enter $7,848.90. Otherwise, multiply the amount on line 8 by the decimal amount on line 9 and enter the result. Also enter this amount on Form 1040, line 48 . **Note:** *Also enter one-half of this amount on Form 1040, line 25.*	**10**	

For Paperwork Reduction Act Notice, see Form 1040 Instructions.   Schedule SE (Form 1040) 1990

Figure 6.3. "Social Security Self-Employment Tax" form.

In addition, many states assess these fees against businesses. They also regulate certain types of occupations, in that the legislatures establish licensing requirements and set renewal fees for these licenses. For example, the state of North Carolina regulates over 30 occupations, ranging from accountants to veterinarians. North Carolina also levies license fees against hundreds of businesses.[4] (Fortunately, writers are *not* among either category!)

Schedule SE (Form 1040) 1990   Attachment Sequence No. **17**   Page **2**

Name of person with **self-employment** income (as shown on Form 1040)	Social security number of person with **self-employment** income ▶

Section B—Long Schedule SE (Before completing, see if you can use the short Schedule SE on the other side (Section A).)

A If you are a minister, member of a religious order, or Christian Science practitioner, AND you filed **Form 4361**, but you had $400 or more of **other** earnings subject to self-employment tax, continue with Part I and check here . . . . ▶ ☐

B If your only income subject to self-employment tax is church employee income and you are not a minister or a member of a religious order, skip lines 1 through 4b. Enter -0- on line 4c and go to line 6a. But **do not** include your church employee income on line 6a.

Part I   Social Security Self-Employment Tax

1	Net farm profit or (loss) from Schedule F (Form 1040), line 36, and farm partnerships, Schedule K-1 (Form 1065), line 15a. (**Note:** *Skip this line if you elect the farm optional method. See requirements in Part II below and in the Instructions.*)	**1**	
2	Net profit or (loss) from Schedule C (Form 1040), line 29, and Schedule K-1 (Form 1065), line 15a (other than farming). See Instructions for other income to report. **Do not** include church employee income from Form W-2 on this line. (**Note:** *Skip this line if you elect the nonfarm optional method. See requirements in Part II below and in the Instructions.*)	**2**	
3	Combine lines 1 and 2. Enter the result	**3**	
4a	If line 3 is more than zero, multiply line 3 by .9235. Otherwise, enter the amount from line 3 here . .	**4a**	
b	If you elected one or both of the optional methods, enter the total of lines 12 and 14 here	**4b**	
c	Combine lines 4a and 4b. If less than $400, **do not** file this schedule; you **do not** owe self-employment tax. (**Exception:** *If less than $400 and you had church employee income, enter -0- and continue.*) . ▶	**4c**	
5	Maximum amount of combined wages and self-employment earnings subject to social security or railroad retirement (tier 1) tax for 1990	**5**	$51,300 ⎸00
6a	Total social security wages and tips (from Form(s) W-2) and railroad retirement compensation (tier 1). **Do not** include Medicare qualified government wages or church employee income on this line	**6a**		
b	Unreported tips subject to social security tax (from Form 4137, line 9) or railroad retirement tax (tier 1)	**6b**		
c	Add lines 6a and 6b. Enter the total	**6c**	
7a	Subtract line 6c from line 5. If zero or less, **do not** file this schedule; you **do not** owe self-employment tax . . . . ▶	**7a**	
b	Enter your church employee income from Form W-2 of $100 or more . .	**7b**	
c	Multiply line 7b by .9235 (if the result is less than $100, enter -0-) . . .	**7c**	
d	Add lines 4c and 7c. Enter the total . . . . ▶	**7d**	
8	Enter the **smaller** of line 7a or line 7d	**8**	
9	Enter your Medicare qualified government wages. See Instructions to see if you must use the worksheet in the Instructions to figure your self-employment tax .	**9**	
10	**Self-employment tax.** If line 8 is $51,300, enter $7,848.90. Otherwise, multiply line 8 by .153 and enter the result. Also enter this amount on Form 1040, line 48	**10**	
	Note: *Also enter one-half of this amount on Form 1040, line 25.*		

Part II   Optional Method To Figure Net Earnings (See "Who Can File Schedule SE" in the Instructions.)

See Instructions for limitations. Generally, you may use this part **only** if:

A Your **gross** farm income[1] was not more than $2,400; **or**

B Your **gross** farm income[1] was more than $2,400 and your **net** farm profits[2] were **less** than $1,733; **or**

C Your **net** nonfarm profits[3] were less than $1,733 and also **less** than two-thirds (⅔) of your **gross** nonfarm income.[4]

11	Maximum income for optional methods	**11**	$1,600 ⎸00
12	**Farm Optional Method**—If you meet test **A** or **B** above, enter the **smaller** of: two-thirds (⅔) of gross farm income[1] **or** $1,600. Also include this amount on line 4b above	**12**	
13	Subtract line 12 from line 11. Enter the result	**13**	
14	**Nonfarm Optional Method**—If you meet test **C** above, enter the **smallest** of: two-thirds (⅔) of gross nonfarm income[4] or $1,600; **or**, if you elected the farm optional method, the amount on line 13. Also include this amount on line 4b above	**14**	

[1]From Schedule F (Form 1040), line 11, and Schedule K-1 (Form 1065), line 15b.   [3]From Schedule C (Form 1040), line 29, and Schedule K-1 (Form 1065), line 15a.
[2]From Schedule F (Form 1040), line 36, and Schedule K-1 (Form 1065), line 15a.   [4]From Schedule C (Form 1040), line 7, and Schedule K-1 (Form 1065), line 15c.

For Paperwork Reduction Act Notice, see Form 1040 Instructions.   Schedule SE (Form 1040) 1990

*U.S. GPO: 1990-265-208

Figure 3 continued.

THE LAW AND THE WRITER

A technical writer's involvement with governments doesn't end as soon as he pays his taxes. He must also contend with other laws that serve to regulate (and, in some cases, protect) him and his work. Foremost among these is the Copyright Act of 1976 — a law that vastly expanded the rights of writers and other "creators", but which also gave them additional responsibilities to protect the creative works of others. In addition, writers are subject to federal (as well as state) laws concerning such matters as libel, invasion of privacy, and trademark infringement. Although the last three areas rarely concern technical writers, they need to at least become familiar with them. For in these matters, ignorance is not bliss.

Copyright: Rights and Responsibilities

Ever since the invention of the printing press, writers have endeavored to protect ("copyright") what they have written to prevent it from being stolen by others who would profit from the hard-earned fruits of their labors. In this, they have had limited success. Realizing this, our forefathers wrote copyright protection into the U.S. Constitution. Article I, Section 8 gives Congress the power to "promote the progress of science and useful arts, by securing for limited times to authors and inventors the exclusive right to their respective writings and discoveries."[5]

The 1790 Copyright Act was passed to implement this article. Since then, several other copyright laws have been enacted, mainly to update previous legislation to keep up with developments in evolving forms of expression and means of communication.[6] Also, in recent years, the Author's Guild and other writer-advocacy organizations have pressured legislators to expand the ownership rights of authors to their works. The latest piece of legislation, the Copyright Act of 1976, is probably the most writer-friendly. Effective January 1, 1978, it applies to all works created after that date. It also lengthens the period of protection given to copyrighted works written before 1/1/78.

Copyright Act of 1976

A creator of a work — writing, painting, song, photograph, film, computer program — owns the copyright on that work as soon as he or she "fixes it in a tangible medium of expression". The medium may be paper (book or article), canvas (painting), magnetic tape or diskette (film, computer program, or musical composition), or anything that renders the creation permanent, perceivable, and reproducible.

Although the particular creative *expression* of ideas can be copyrighted, the ideas themselves cannot. For instance, if you want to write about the General Theory of Relativity, you don't need to obtain copyright permission from the trustees of Albert Einstein's estate. That is because Relativity is a *concept* (albeit a revolutionary one), not a unique creation. It belongs to all of mankind — a collective ownership that, I think, Einstein would've approved of. However, that fact wouldn't give you the right to reproduce substantial portions of a copyrighted book or paper written by Dr. Einstein in which he expounds upon his famous theory and attribute these portions to yourself. Doing so would be *copyright infringement* because you would be misappropriating a particular form of expression of this idea, not just the idea itself.

Nor does copyright protection extend to research and other collections of facts and data. Protection only applies to the creative work of an author. The authors of the act believed that giving copyright protection to facts and data would serve to inhibit research activity.

What kinds of rights does a copyright owner have? The Copyright Act lists five exclusive rights, of which only three apply to technical writings. These are the rights to:

- Reproduce the copyrighted work in copies
- Prepare "derivative works" based upon the copyright (These include such works as translations, condensations, sound recordings, or any other form in which the work may be "recast, transformed, or adapted".)
- Distribute copies of the copyrighted work

In other words, these are the rights which give the copyright owner access to any revenues derived from the work. However, these rights do not necessarily give the copyright owner *all* of the revenues. If she contracted with a firm to publish her book, she would receive that portion of sales revenues her contract specified. Never would this portion be 100%. And if the publisher sold the translation or other subsidiary rights to her book, she probably would receive just a share of the proceeds, not all of them. Nonetheless, copyright ownership is valuable, if for no other reason than it allows her the freedom to sell or retain these rights, especially those concerning derivative works. Without copyright ownership, she would have little or no control over the disposition of her work.

A work is protected by copyright for a term equal to the lifetime of the creator plus 50 years. If there are several authors, this term applies to the last surviving author. This provision applies to most works published (or in the process of being published) on or after January 1, 1978. The exceptions are anonymous or pseudonymous works and works "made for hire" (discussed below). For these, the copyright is valid for 75 years from the year of first publication or 100 years after their creation, whichever comes first.

Exceptions to the 1976 Act

What about works published before January 1, 1978? These were covered by the old copyright law, which provided two 28-year terms of protection. However, to renew his copyright, the owner had to apply for a renewal between the 27th and 28th year of the first term. The 1976 law extended the second term to 47 years , giving old copyrights a total duration of 75 years. It also extended the renewal application period to December 31 of the calendar year in which the first term would have expired.

Any work whose copyright has expired or which was never copyrighted in the first place is said to be in the "public domain". That means that anyone can use it for whatever legal purpose he or she desires. Moreover, once a work falls into the public domain, it can never be copyrighted. For instance, the writings of Edgar Allan Poe (who died in 1849) would be in the public domain. However, if an author published an anthology of his short stories, in which the stories were arranged in a certain way and were illustrated with certain "grotesque and arabesque" drawings, then the anthology would be copyrightable, though the stories themselves wouldn't be.

By their very nature, certain works are always in the public domain. These include works of the U.S. Government, as well as statements by any of its representatives of the executive, legislative, and judicial branches. Presumably, works prepared by individuals or firms under contract or grant to the government would also be in the public domain, though the Act does not address them explicitly. The key consideration is whether these works are merely prepared by contractors/grantees in lieu of government employees who could prepare them just as well themselves.

"Works for hire" comprise another category of works that cannot be copyrighted by their authors. The Copyright Act defines two kinds of "works for hire": (1) a work prepared by the employee for his employer in the course of his employment and (2) a work "specially ordered or commissioned for use as a contribution to a collective work...as a translation, supplementary work, compilation, instructional text,...or as an atlas". Full-length books may *not* be works for hire unless they are instructional texts. For this second category of works, the "for hire" status must be confirmed in writing by the author and employer. Preferably, the instrument confirming the arrangement would state expressly that the work is, in fact, a "work for hire". The copyright for a "work for hire" is held by the employer, not the author.[7]

Is a technical article or book a "work for hire"? Almost always, yes. In the case of a journal article, the copyright is held by the periodical, which typically copyrights the entire contents of every issue. However, there are exceptions. For instance, if an author wanted a journal to publish an excerpt from his soon-to-be published book, he or (more likely) his publisher would grant the journal only a one-time right to publish the piece.

As mentioned in Chapter 4, the copyright to a technical book is invariably owned by the publisher, not the author. That is standard practice. In terms of the 1976 Copyright Act, the publisher has "commissioned" the author to write his book, which is, in effect, a "work for hire". But how can this be, if the Act specifically excludes full-length books from the works for hire category? The way this is done is to designate these books (explicitly or implicitly) as "instructional texts". Come to think of it, many (if not most) technical books would fall into this category. If you think that the book you've written *doesn't*, then you should have a serious chat with your publisher. It's quite possible that, despite what your contract says, *you*, not he, own the copyright on your book.

Although a work is legally copyrighted as soon as it is created in a tangible form, *proving* that requires some effort. The best way to establish copyright is to register it with the U.S. Copyright Office [Library of Congress, Washington, D.C. 20559, (202)-479-0700]. This requires completing a form (Form TX) and sending it to the Copyright Office along with a $10 registration fee and a copy of the work, if unpublished (two copies, if it's already published).

If you choose not to register your work and it gets published with a "copyright notice" affixed to it, the Act still requires you to deposit it with the Copyright Office. This must be done within 3 months of publication. (In most cases, the publisher will take care of this, either for you or himself, depending upon in whose name the copyright is registered.) Copyright notice consists of three elements: (1) the word "copyright" or the symbol ©, (2) the author's full legal name, and (3) the year of first publication. (Example: "Copyright 1991 William M. Vatavuk.") This notice must be affixed to the title page of the work or anywhere else where it would be easy to find. However, for works created after March 1, 1989, copyright notice is *not* required to establish copyright, though creators are nonetheless encouraged to use it.[6]

The 1976 Copyright Act protects works inside the U.S., but what about outside its borders? Fortunately, it would have the same sort of copyright protection in most countries in the world, thanks to the Berne Convention, an international copyright agreement that the U.S. (finally) signed in 1988.[6] International copyright protection is no triviality. It can protect both the author and publisher from the loss of thousands of overseas sales dollars. For this reason, it has been a burning issue since the days of Poe and Charles Dickens, both of whom lobbied for international copyright agreements. In those days, it was common practice for U.S. and British publishers' representatives to gather at docks to await the arrival of overseas vessels. When the passengers debarked, these representatives would pay them exorbitant prices for any popular foreign books they had taken with them to read during the long ocean voyage. The publishers would then take these books, remove the covers, reprint the contents, and market them under their own imprints. Needless to say, the original publishers wouldn't reap a penny from the sale of these pirated editions.

Plagiarism: The Worst Sin Words Can Describe

There are several ways one may infringe upon a copyright. Reproducing a copyrighted work and then selling it is perhaps the most obvious infringement, as it enriches the thief at the expense of the copyright owner. However, there is another type of infringement, one which may not increase the thief's net worth, but which very well may enhance his professional stature. And that, fellow professionals, is a sin called *plagiarism*. What is plagiarism? Lee Wilson, a noted intellectual property lawyer, defines plagiarism as "taking the 'fundamental substance' of another's work" and attributing it to oneself.[6] However, plagiarism is only one of three kinds of copyright infringement based on the notion of "substantial similarity". This notion says, in essence, that a work infringes upon another copyrighted work if it is "substantially similar" to it. Word-for-word copying is a form of this infringement, but so is writing a work that has "comprehensive literal similarity" (i.e., which reproduces the essence or "theme" of the copyrighted work).

A third type of infringement is taking of "portions of a work which are important to the impact and character of the work...but which do not amount to a large portion of the infringing work."[6] For instance, if you took a figure from a book containing hundreds of figures and this figure showed information that synopsized data on most of the other figures or was otherwise important to the "impact and character" of the book, then you would be infringing the copyright.

The key word here, of course, is "copyrighted", as you may legally borrow the words of another if they are not copyrighted. In fact, in addition to government publications, the Copyright Act does not extend copyright protection to several other works, including names of products, services, or businesses; pseudonyms; slogans; blank forms; measuring and computing devices; and compilations of commonly available information that contain no original material (e.g., calendars).[6]

However, as I discussed above, if the work is in writing or has been otherwise expressed in a tangible form, it *is* copyrighted, regardless of whether it has been formally registered. This means that you may not "borrow" the rough, handwritten draft of your colleague's research paper, make a clandestine photocopy, and then include portions of it in a paper you're writing. Doing so would be a violation of your colleague's copyright. In such a case, he could have solid grounds for an infringement lawsuit.

It would also be an *unethical* act. As any professional worth his or her pay knows, plagiarism is a very nasty crime. Plagiarists are *persona non grata* in the academic community. A student caught plagiarizing likely would be expelled from a college or university, while a professor (tenured or otherwise) who plagiarized could find himself severely reprimanded or removed. Their reputations would be shattered. And this punishment would be in addition to any redress available through the courts. In fact, whether the stolen material was copyrighted probably would make no difference to the university.

A similar fate could befall a plagiarist in private industry, government, or another institution where research and publication are stocks in trade. The punishments could be less severe, as ideas are no more important to any group of people than they are to academics. Still, plagiarists would, at the very least, suffer irreparable damage to their personal and professional reputations. If they're lucky, they may be allowed to keep their jobs. But they may find future avenues of research and publication roadblocked, for their colleagues would be extremely reluctant to work with them or give them access to their findings.

Since plagiarism is so vile (the very word evokes disgust in many), professionals have learned from their formative years that they must give proper attribution to any material they use in a work, regardless of how widely that work would be disseminated. (In fact, to many ethicists, whether the work was eventually published or not wouldn't matter. The crime would be just as venal in either case.) Both directly quoted and indirectly quoted (paraphrased) material must be attributed. Attributing direct quotations is straightforward: you simply enclose the words in quotation marks and add a reference or footnote number to key it to the full citation of the source in the list of references. If you delete words from the original quotation, so indicate with ellipses (...). Similarly, if you add words for clarification, place them within brackets ([ ]) to indicate that they weren't part of the original quote. Any writer's style manual would tell you to do likewise.

> Example: "If Sir Isaac Newton had actually been sitting under an apple tree when the proverbial apple dropped on his head, he undoubtedly would be too busy moaning and rubbing the bump on his noggin...[than to] ponder the mysteries of gravity."[77] (77. Pokefun, Peter, "The Levity of Gravity," *Satirists' Semi-Annual,* Summer 1991, p. 47.)

Paraphrasing isn't as easy to handle. This is so because it isn't always easy to determine whether what you've cited is a paraphrase of someone else's thoughts or just an expression of your own. Consider the previous example. If you wrote, "I doubt that Sir Isaac Newton really discovered gravity just because he was hit by a falling apple while sitting under a tree one day," then you *might* be paraphrasing Mr. Pokefun. Then again, you simply might be offering an opinion — a commonly held one, incidentally — that this story of Newton's deducing the Law of Universal Gravitation as a result of being clobbered by a falling winesap is just that: a story. It has as much to do with the history of science as a certain president's felling of a cherry tree has to do with American history.

However, if Mr. Pokefun's article had contained more specifics about the apple, the apple tree, the time of day, and other pseudoscientific parameters, and then you had rearranged those data and included them in your writing, it would look more like a paraphrase than an original contribution to the wonderful world of satire. Consider this quote from Mr. Pokefun's piece:

> Now, Sir Isaac *could* have carefully measured the following parameters: (1) distance of overhanging tree limb from ground, (2) wind speed and direction, (3) air temperature and pressure, and (4) mass of falling apple. With these data in hand, he *could* have developed the rudiments of gravitational theory. Of course, for his theory to be validated, he would have had to repeat his falling apple experiment many times. However, we doubt whether: (a) the tree in question bore enough apples and (b) Sir Isaac's egg-shaped noggin could withstand the punishment.

Upon rearranging a few phrases, one might obtain the following paraphrase of this passage:

> As a scientist, Newton could have measured the mass of the falling apple, the distance of the tree limb from the ground, the wind speed and direction, and the air temperature and pressure. From these figures, he could've developed a crude theory of gravitation. But he'd have had to replicate the experiment many times to validate his theory. The supply of apples was limited, as was his tolerance for cranial blows.[78]

Notice that I've given this second passage a citation, as it is clearly a paraphrase. If I hadn't cited it, a disinterested person might conclude that I'd lifted it from Mr. Pokefun. Such a conclusion would be well founded because the content and, to some extent, the form of the paraphrase closely resemble the passage in Pokefun's article.

Most professionals sincerely believe that if they properly attribute a quote, paraphrase, or other information to a source, they cannot be accused of infringing the copyright. In most cases, that's true. But if you should appropriate a large and/or essential amount of material and use it in your work, you may be infringing, *even if* you cite the source. In other words, citing the source gets you out of the plagiarism frying pan, but could land you in the infringement fire. This leads us to two more questions: (1) what constitutes a "large/essential amount of cited material"? and (2) how much material may you use (with proper citation to the source) without infringing the copyright? These and like questions are central to the doctrine of "fair use".

The Copyright Act defines fair use as the use of a copyrighted work for such purposes as "criticism, comment, news reporting, teaching (including multiple copies for classroom use), scholarship, or research..."[7] Nonetheless, determining what is or isn't fair use is rarely a cut-and-dried task. Each situation must be evaluated individually. However, to make this evaluation easier, the 1976 Act provided four factors:[7]

1. The purpose and character of the use, including whether such use is of a commercial nature or is for educational purposes
2. The nature of the copyrighted work
3. The amount and substantiality of the portion used in relation to the copyrighted work as a whole
4. The effect of the use upon the potential market for or value of the copyrighted work

The first two factors are more or less self-explanatory. Factor #3 is probably the one that puzzles writers most. That is, a writer might wonder whether the portion that he is quoting or paraphrasing is "substantial" compared to the whole of the work. Most of the time, common sense will provide the answers. But in other cases, such as the one I describe below concerning the handbook tables, it may not. It is definitely a "judgment call".

The fourth factor is one which the courts often consider the most important.[7] A writer might extract a substantial portion of a work for use in a second, commercially oriented work, but if this extraction would have little or no detrimental financial effect on the original work, then a court might very well deem it to be "fair use". Several court cases have reinforced this interpretation of the doctrine.

Since a fair use determination is often so difficult to make, it behooves you, as a writer, to err on the side of caution. If you believe, or even suspect, that the material you've extracted from a work constitutes a substantial portion, then contact the copyright owner (publisher, author, or both) to request permission to reprint the material in question. With your request, enclose two signed copies of a reprint permission form (see Chapter 4 for a sample) and a copy of the material you wish to reprint. Ask the copyright owner to sign one copy of the form and return it to you, while keeping the other copy for his/her records. As an added courtesy, also enclose a stamped, self-addressed envelope.

I daresay that, in most cases, the copyright owner will approve your request to use the material, provided that you use an appropriate citation (e.g., "Reprinted by permission of Shade Tree Publishing, Inc."). However, in other situations, the copyright owner may approve your request, but charge you a fee (royalty) for using the information, usually for one time only. The amount charged will vary according to the length of the excerpted material and the owner's appraisal of its value. Royalties of $10 to $20 per printed page are typical, in my experience.

You will have to decide whether the material is worth the price.

Obtaining these permissions isn't as easy as it may seem. Not long ago, we tried to get permission to reprint certain data we had obtained from two handbooks, both copyrighted by the same publisher. We had extracted these data from several tables in these handbooks and compiled them into a single, composite table. We had planned to include the latter table in a federal government report. Thinking that this report would be published by the government, which would not profit from its sale, we naively assumed that the publisher would approve our request forthwith. How wrong we were! The publisher required us to submit not only a copy of the composite table, but also a copy of *each* table in the two handbooks from which we had extracted the data. This amounted to six tables which occupied dozens of pages of text. After we sent these tables to the publisher, he thereupon told us that we could use the material in our report (the good news), but that they would charge us $15 per *handbook* table to use it (the bad). This unexpected turn of events bowled us over. We grumbled among ourselves but, in the end, decided to pay the fee anyway. Nonetheless, we felt that we had been had. After all, compared to the hundreds of tables and thousands of pieces of data in the two handbooks, the amount of information in the composite table was minuscule. If publishing this table didn't constitute "fair use", I don't know what would have.

However, let's take a moment to consider the publisher's side of the story. From his perspective, material taken from his copyrighted works and published in a government report would be more vulnerable than material published in a commercial work. Why? Simply because a government publication cannot be copyrighted. It is public domain material that anyone may republish and sell. (In fact, a prominent technical publisher earns much of his revenues by reprinting and selling federal government reports.) Now, this does not relieve a republisher of the obligation to make sure that none of the material in the government report infringes a copyright. But, in reality, *proving* this second-hand infringement is not easy, especially since the government would be a party to any such suit. Publishers know this full well. Hence, they are less willing to allow the government to reprint portions of their copyrighted works than they would to someone in the private domain.

Compared to another situation I encountered, this table was a bargain. A few years ago, I planned to write an article on a famous 19th century keelboatman for a state government-published popular history magazine. To begin this article, I thought I'd reprint a few stanzas of a

---

* I know, I know. I'm being vague about these two cases. But I'm intentionally muddying the water because I don't want the parties involved to become annoyed. Besides, the names aren't important to this anecdote.

song written by a major Hollywood studio for a movie about this legendary hero.* When I asked the studio for permission to use the lyrics, they agreed to sell me "one-time reprint rights" to them for an astronomical sum — about $2000, I recall. I told the studio that this was about five times the amount that was going to be paid for the article. "Besides," I pleaded, "this is a publicly owned and operated magazine. No one makes any money from it." They nodded their corporate heads sagely, then asked me to give them certain information about the magazine — circulation, readership, contents, frequency of publication, etc. After obtaining these facts and conveying them to the studio, they said they'd review them and contact me soon. I thought that they might reconsider and let me reprint the lyrics for free. Not a chance. A few weeks later (things move slowly in large Hollywood studios), they sent me a letter in which they "compromised" by lowering their reprint fee to only (!) $200. At that point, I just gave up and deleted the lyrics. The words were simply too *golden* for my pocketbook.

However, in a few cases, the copyright owner may deny a reprint request outright, leaving you with the choice of either deleting the material or including it. If you do include it, you'll be taking the chance that the owner will never notice that it's been appropriated. After all, given that over 50,000 books are published a year, what would be the odds of the owner's even noticing the book, let alone the "borrowed" material? And even if the owner did notice it, what would be the odds of his filing an infringement lawsuit?

This sort of rationalizing can lull us into a false sense of security, thinking that we are protected by the "information explosion" and the copyright owner's reluctance to sue. But don't count on it. Though it is more troublesome in the short run, in the long haul it is better not to appropriate copyrighted material than to risk legal action. Defending a lawsuit is an expensive, nerve-wracking, and reputation-wrecking experience. And that's if you *win*. If you lose, you can lose…just about everything. For courts can, and do, award huge judgments to plaintiffs who win infringement suits. These judgments may include not only "compensatory" damages, but also "punitive" damages, the latter being meant to "punish" the defendant for his wrongdoing. To add insult to injury, the court may order the defendant to pay the plaintiff's attorneys' fees, which, at $100 to $300 per hour and more, can be as high as the damages. Reference 6 provides a first-hand glimpse of the litigation process, how it works, and how it could affect you if you were on either the sending or receiving end of a lawsuit.

Libel and Other Crimes of Dispassion

Speaking of suits, there are other kinds of legal actions that can be brought against a writer who fails to mind her Ps and Qs. They are (1) libel, (2) invasion of privacy, and (3) trademark infringement. Fortunately, the first two matters rarely arise with technical writing, mainly because this kind of writing usually centers on thoughts or things (or both), but rarely on people. Thus, the odds of your libeling persons or invading their privacy in the course of your writing would be rather long.

Still, we can envision situations where these issues *could* arise. For instance, you could have written a biography of a Nobel Prize-winning scientist that is "unauthorized" (i.e., the subject hasn't given permission for you to write the book). And, in this book, you might have inadvertently made some disparaging remark about the scientist — one so disparaging that she decides to sue you and your publisher. If you knew that this remark might be considered libelous, you wouldn't have included it in the first place.

Libel

The first thing to learn about libel, invasion of privacy, or any of these torts is that they are *not* well defined. Unlike scientific laws, which are immutable, legal principles have evolved over time. Subject to political pressures, changing social mores, and other forces, the law changes shape relentlessly. In our legal system, the law has two major constituents: legislation and case law, the latter consisting of all of the court decisions (state and federal) rendered since the U.S. Constitution was ratified. Indeed, in some matters, case law ("precedent") may have more bearing on the outcome of a lawsuit than the governing legislation.

Libel is a good example. Rooted in British common law, libel law has evolved in this country from colonial times, when it was both a civil and a criminal act to defame public authorities, to our present-day litigious society. Two landmark cases have paved the way. The first, and arguably most momentous, libel suit in this nation involved John Peter Zenger, a New York editor and publisher of the *Weekly Journal*, who was jailed in 1734 for criticizing the policies of the British government. In 1735, he was tried for "seditious libel" by a jury and found not guilty, thanks primarily to an eloquent defense by his attorney, one Andrew Hamilton.[5]

Zenger's victory established the right of the press to freely criticize public officials without fear of retribution. Over time, the courts have reinforced and expanded the media's First Amendment rights. The second landmark decision, *New York Times Co.* vs. *Sullivan*, was rendered by the U.S. Supreme Court in 1964. In this case, L.B. Sullivan, a Montgomery, Alabama city commissioner, sued the *New York Times* for libel, alleging that his reputation was damaged by an advertisement placed in that newspaper by a group of civil rights supporters. Sullivan won a libel judgment in the lower (Alabama) courts, but the Supreme

Court overturned the decision. In its majority opinion, the Court threw out the Alabama libel law (and that of many other states). In its place, it substituted what become known as the "actual malice" doctrine. In essence, this meant that for a public official, such as Sullivan, to win a libel action, he "...had to prove not only that the [libelous] statements were false, but also prove convincingly that the press published them while knowing them to be false or with reckless disregard to whether they were false or not."[8]

As we might expect, it is difficult to prove actual malice, as establishing proof requires the plaintiff virtually to read the defendant's mind and ascertain his/her intentions when making the criticisms. Nonetheless, since the *Times* decision, the press has lost several well-publicized libel cases brought against them by public figures. Among them was comedienne Carol Burnett's successful 1983 suit against the *National Enquirer* magazine for having printed a false and malicious story about her.[8] These and other decisions prove that, although the press' right to self-expression is guaranteed, this right does not extend to libeling public figures.

Keep in mind that these cases concerned *public* figures — celebrities, elected officials, and others in the public eye. However, when it comes to *private* figures — most of the rest of us — libel law is much less forgiving. It is said that "the right to be left alone" is the most basic of all human rights, and our courts have agreed. (The infamous *Roe* vs. *Wade* abortion decision is a classic illustration.) Hence, the courts have been harder on defendants sued for libeling private individuals.*

But what *is* libel, exactly? In general, libel is "a false statement which is 'of and concerning' the plaintiff, made as a statement of fact...which causes harm to the plaintiff by injuring her or his reputation or subjecting her or him to shame and ridicule in the community, and which is the result of some omission or fault of the defendant."[6] Harmful statements include (but may not be limited to) false remarks about a person's health (whether he has a "loathesome disease"), mental competence, chastity (these days, a euphemism for "sexual habits and orientation"), ethics, honesty, sobriety, criminal record, and professional competence. Anything disseminated through the media about an individual that falls into one or more of these categories is potentially libelous and grounds for a suit. (I say "potentially" because these categories are just guidelines. The ultimate determination of whether a statement is libelous will made by a court of law.)

Technical writers should avoid these "libel traps" at all costs whenever they refer to other living persons in their writing. The operative

---

* This distinction between "private" and "public" individuals is so crucial that, in some libel cases, one of the first things the defendant's attorneys do is try to establish that the plaintiff is a public figure. If they succeed, they've gained a significant advantage.

word here is "living", as by law, one cannot libel a dead person. In other words, because a person has died his or her reputation cannot be damaged, since it no longer exists. What about the person's descendents? Couldn't they be harmed if their deceased ancestor's name were dragged through the mud? Possibly, but that would be beside the point. The point would be that the inflammatory statement(s) would be directed against the ancestor, not his family. In the court's eyes, the ancestor would be where nothing or no one could harm him or her.

Another defense against a libel action is the truth. If what you have written about a person is unquestionably true, then it is not libelous. Suppose you wrote: "Professor Slipshod is a plagiarist. In 1989, he was discharged from Makeshift University, a prestigious private institution, after having taken a colleague's research results and published them under his own name in the October 1987 *Journal of Applied Marginalia*." If this were true, and Slipshod sued, you'd probably win. But you would've had to prove these bold statements. Presumably, there would be a record of Slipshod's firing. And, of course, you could obtain the issue of the journal containing his pirated paper. In any event, you would be foolish not to obtain the necessary evidence before these two sentences got into print.

Invasion of Privacy

Even if you "had the goods" on Professor Slipshod, he could still take you to court and win. How? By suing you for *invasion of privacy*. Recall that Makeshift U. is a private school. That means that Professor Slipshod is a private employee. By revealing his past indiscretions, you would have violated his privacy rights. More than likely, Slipshod was discharged very quietly. Officially, no one but the adminstration and faculty were aware of the incident. Unofficially, the entire student body and half the population of Makeshift, N.C., knew of it as well, as the grapevine in any college town is very efficient. (Come to think of it, that's how *you* found out about the firing in the first place. You confirmed the rumor by questioning a friend of yours in the chancellor's office. Which leads us to another question: is your friend a reliable source — reliable enough to give credible testimony in court?)

In this instance, you had invaded Slipshod's privacy by making a public disclosure of a private fact. This is just one of several forms of invasion of privacy: Another is "false light invasion of privacy". According to Wilson, this sort doesn't harm a person's reputation, per se, but it does portray him or her falsely to the public, injuring the person's dignity and causing mental or emotional distress.[6] If you photographed Dr. Pulmonary as she stepped out of a tavern and then ran this photo with the caption, "Area physician enjoys a night on the town", she could sue you for false light invasion of privacy. Why? Nightclubbing is neither illegal nor immoral. But Dr. Pulmonary happens to be a recovering alcoholic and, moreover, the state president of Alcoholics Anonymous. To her and her associates such a photo would be embarrassing, to say the least.

There are two more kinds of invasion of privacy. The first is *intrusion,* whereby you either physically trespass on a person's property or gain access to it under false pretenses. You can also photograph someone, eavesdrop electronically, or use some other clandestine means to intrude upon a person's God-given right to be left alone. Dr. Pulmonary could sue on these grounds, as the infamous barhopping photograph was taken surreptitiously and without her consent.

Finally, suppose that Professor Slipshod's colleague — the one he had plagiarized — was Dr. Quark, world famous theoretical physicist. You may recall that he earned his fame and modest fortune from a highly rated TV series during which he revealed, in everyday language, the mysteries of relativity and quantum physics. Suppose further that you had taken a file photo of Dr. Quark and published it with an advertisement for electronic calculators. The photo caption read: "A 'Presto' calculator lights up this man's numbers. Shouldn't it light up yours?" Nothing defamatory here. Besides, another friend at Makeshift U. told you that Dr. Quark always used Presto calculators. In fact, he carries one with him wherever he goes. So where's the problem? The problem is, you used this photo of Quark for commercial purposes *without* his permission. This is known as *misappropriation invasion of privacy.*[6] Since Quark is a public figure whose name is "bankable", he could bring suit on the grounds that you used his likeness without compensating him for the endorsement — an endorsement that he gladly would have given (if the price were right).

Trademarks

As mentioned above, ibel and invasion of privacy are rarely areas of concern to technical writers, as their writings seldom involve individuals. Trademarks are another matter, however. Often we do find ourselves mentioning brand name products in our technical articles and books. We may not wish to mention them, but we often have little choice. For instance, suppose that you were writing an article on the dry photocopying process for a technical journal or, for that matter, a popular science magazine. At some point, you would have to mention Xerox Corporation and their pioneering efforts in this field. You also likely would write about this company's photocopiers and how they have evolved over the past quarter century. You might even mention them by name and model number. But if you do, you'd have to indicate that the model name is a *registered trademark.* The best and most common way to indicate this is via the letters "TM" or "®" (e.g., XEROX Model XYZ ®). Moreover, if you discussed the photocopying process in this article, you also would be certain to call it "photocopying", not "[this corporation's name] + ing". Xerox Corporation and many other developers and manufacturers of unique, ubiquitous products are understandably upset when the media misappropriate their product names, not so much for commercial gain, but merely as a shorthand substitute for proper English. Consider these household names: "nylon", "escalator", "cornflakes", "kerosene", and "yo-yo". Each was once a registered trademark that lost its protection due to its becoming "generic", meaning that the public came to use the trademark to describe the product or service that it represents.[6] Incidentally, not only may product and process names be registered trademarks. So may symbols, slogans, designs, and common words used as product names (e.g., ACE® bandages).[6]

When writing, if you are not certain whether a product is or isn't a registered trademark, you can do one of three things: (1) delete the word and write around it, (2) contact the firm that manufactures the product to see if the name is registered, or (3) pore through such references as *The Trademark Register of the United States*, *The Trademark Design Register of the United States*, or *The Compu-Mark Directory of U.S. Trademarks*.[6] These directories list all registered trademarks. Finally, when in doubt, drop in the "circle R". It doesn't pay to risk an infringement suit.

This concludes the "how-to" portion of this book. Up to now, you've read about why and what you should write and how to go about getting it published as an article or book. You've also gotten a peek at today's publishing market and some valuable (I hope!) advice on how to cope with taxes, copyrights, libel, and other vexations. If all you've wanted to get out of this book is self-help information, you have my permission to stop reading right now and get to work — writing, that is. However, if you'd like to read the interesting things some leading writers, editors, and publishers have had to say about their trade, stick around. There's more to come.

REFERENCES

1. 1990 Form 1040. (Washington, D.C.: U.S. Department of the Treasury, Internal Revenue Service, 1989).
2. *Tax Guide for Small Businesses*, Publication 334 (Washington, D.C.: U.S. Department of the Treasury, Internal Revenue Service, 1990).
3. *Business Use of Your Home*, Publication 587 (Washington, D.C.: U.S. Department of the Treasury, Internal Revenue Service, 1990).

4.   *Directory of North Carolina State Business Licenses and Permits* (Raleigh, N.C.: Department of the Secretary of State, Business License Information Office, 1990).

5.   Hicks, J. D., et al. *The Federal Union,* 4th ed. (Boston: Houghton Mifflin Company, 1964).

6.   Wilson, L. *Make It Legal: A Guide to Copyright, Trademark, and Libel Law; Privacy and Publicity Rights and False Advertising Law* (New York: Allworth Press, 1990).

7.   Norwick, K. P. *The Rights of Authors and Artists* (New York: Bantam Books, 1984).

8.   Dill, B. *The Journalist's Handbook on Libel and Privacy* (New York: The Free Press, 1986).

7 Technical Writers, Editors, and Publishers Speak Their Minds

Nothing succeeds like success
(Proverb)

The above adage applies to technical writing, as well as to any other endeavor. As an endeavor, or business if you will, technical writing involves thousands of professionals. There are the writers, of course, but also the editors, publishers, and sundry support personnel that keep publishers and periodicals operating in the black. Other than direct observation and participation, the best way to learn about the technical writing business is to talk to those involved in it, especially the more successful among them.

With this in mind, I wrote several leading technical writers, editors, and publishers to ask if they'd like to share their thoughts and opinions about their respective trades. I enclosed a brief questionnaire that contained questions purposely designed not to elicit specific responses, but to stimulate their thinking. Since the interests, objectives, and viewpoints of writers differ somewhat from those of editors and publishers, I developed a different questionnaire for each group. These are shown in Figures 7.1 and 7.2, respectively. I'm pleased to say that their enlightening, thought-provoking responses offer fascinating insights into their work and, what's more, *themselves*.

The respondents include three each from the writer and editor-publisher professions. Of the writers, two are full-time freelancers, while the third is a university professor. The editor-publishers include the editor of a technical periodical, the editor-publisher of another periodical, and the vice-president/editor-in-chief of a book publishing firm. For these three, the job titles are somewhat misleading. Though

Please answer the following questions as completely as you like (handwrite or type). Feel free to ad lib as you see fit. If you run out of space, just continue on a separate sheet of paper.

1. Would you please provide some biographical information. (If you have a prepared author bio, feel free to attach it.)

2. Tell us a little about your writing career. What was your first published technical work? List your most significant technical publications. Which of these has meant the most to you and your readers? Why?

3. As a successful technical writer, you've learned what it takes to get into print. Would you give us some suggestions on how to write well and get this writing published? For example, would you recommend any reference books to would-be writers?

4. Writing not only requires knowing what to write, how to write, and where to sell it. It also requires a healthy dose of *self-discipline*. As we know, discipline and good habits are close relatives. Would you please describe your writing habits?

5. Many technical professionals harbor the misconception that they cannot write. For this reason, they avoid it like the proverbial plague. What do you think about this misconception?

6. If you could summarize your "philosophy of writing" in one or two sentences, what would you say?

Last. Do you have anything to add about writing, publishing, publishers, or the writing business in general? Do you think I've left anything out here? Please be frank.

Figure 7.1. Author questionnaire. Thoughts from a successful technical writer.

nominally editors, they do very little editing, as most of us envision the work. That is, they rarely review manuscripts, revise the wording of text, lay out pages, etc. These "copy" editing chores are usually left to subordinates. One of the three does some "acquisition" editing — acquiring manuscripts from writers for publication; but the other two do little or none of this work. In fact, they function more like publishers

Please answer the following questions as completely as you like (handwrite or type). Feel free to ad lib as you see fit. If you run out of space, just continue on a separate sheet of paper.

1. Would you please provide some biographical information. (If you have a prepared bio, feel free to attach it.) Also, please attach a black-and-white photo, suitable for publication.

2. What is your overall assessment of the technical publishing business today? Do you think it is healthy? weak? growing? declining? For instance, do you think it is stronger financially than trade publishing? If so, why?

3. As a successful technical editor, you've learned what it takes to get into print. Would you give readers some suggestions on how to write well and get this writing published. For example, would you recommend any reference books to would-be writers?

4. Writing not only requires knowing what to write, how to write, and where to sell it. It also requires a healthy dose of *self-discipline*. So does editing. In what ways do you think an editor's and an author's "work habits" would compare or contrast?

5. Many technical professionals harbor the misconception that they cannot write. For this reason, they avoid it like the proverbial plague. What do you think about this misconception?

6. If you could summarize your "philosophy of editing" in one or two sentences, what would you say?

Last. Do you think I've left anything out here? Please be frank.

Figure 7.2. Editor/publisher questionnaire. Thoughts from a successful technical editor/publisher.

than editors, in that much of their daily work is involved with the business aspects of publishing — preparing budgets, hiring and firing, dealing with suppliers, printers, distributors, and so forth.

These six men reflect a mix of education, experience, and expertise. However, they have one thing in common: a dedication to technical writing that has borne fruit, both professionally and personally.

THE WRITERS

Robert M. Bly

Bob Bly disproves the adage, "Engineers can't write." For Bly not only writes and writes well, but for many years he has earned a good living from the wordsmith's trade. A chemical engineering graduate of Rensselaer Polytechnic Institute, he worked in industry in that capacity for several years. Eventually, he came to specialize in writing technical advertisements for equipment, materials, and other products. One day, he decided to strike out on his own as a freelance ad copywriter. Since then, he has written 19 books and hundreds of articles, developed and presented seminars to aspiring writers, and prepared audio cassette programs from them. His books include *Technical Writing: Structure, Standards, and Style* (McGraw-Hill), his first, and *The Copywriter's Handbook* (Henry Holt); but his best-known work is *Secrets of a Freelance Writer* (Dodd, Mead), a book, says Bly, that has helped thousands of part-time and full-time freelance writers make more money." Ironically, the writing that has earned him the most money and has advanced his career most steadily has been work that does *not* carry his byline. These include ads, sales letters, press releases, direct mail packages, and catalogs, all written for private clients. As he describes in *Secrets of a Freelance Writer,* this work can be quite lucrative. However, because his writing has been so well received by his clients, he can demand and get high fees. These diverse clients include Sony, Grumman, GE Solid State, Philadelphia National Bank, and Value Rent-a-Car. In addition to writing consistently well, Bly has strived to conduct himself as a *business professional,* a demeanor that many freelancers eschew. "Too many writers act like writers, not business professionals, and this doesn't impress clients," he writes in *Secrets of a Freelance Writer.* "[Y]ou will get your best results with a client by acting as a listener, counselor, problem solver, and business consultant — not an *artiste.*"

Good advice, indeed. Bly's views on other writing matters are equally sensible.

On achieving writing success. "The key to getting published is not writing ability but marketing — targeting the right material to the right editor and the right audience. If you can offer an editor or publisher useful knowledge of a specific technical field clearly presented, you won't have to write like a literary genius to get published."

On writing work habits: "I do my client work (reading, writing, and consulting with my clients) in the mornings; afternoons are for routine office work and for writing books and articles under my own byline."

About the technical professionals-can't-write syndrome: "I don't think it's a misconception. Many technical professionals *are* lousy writers. Why should this be shameful? Many writers are lousy at technical things." (his emphasis)

Bly's "philosophy of writing": "The keys to good nonfiction writing:

1. Present information useful to the reader.
2. Appeal to the reader's self-interest. Write what interests them, not what interests you.
3. Write about what you know. Knowledge gained through real-life practical experience is what readers want, not theory.
4. Write in a clear, friendly, conversational tone, like a patient teacher looking over the reader's shoulder."

Roy Meador

Like Bly, Roy Meador is a full-time freelancer. Like Bly, he views writing differently from the way most of us do. His viewpoints are those of a man who earns his living solely from the words he puts on paper.

A native Oklahoman, Meador earned his B.A. and M.A. degrees at the University of Southern California and Columbia University, respectively. Following service as a Naval officer, he worked as a technical, promotional, and advertising writer in New York City and, since 1972, in Ann Arbor, Michigan. His writing experience includes preparation of books, proposals, technical reports, manuals, advertising, annual reports, and other written products for university, company, and association clients. His clients have included General Motors, the U.S. Department of Energy, and the State of Michigan.

In 1975, he published a biography focused on the scientific contributions of Benjamin Franklin. His other books concern future energies, modern technology, cogeneration, and district heating, and the preparation of proposals. Re: the last, his 1985 book *Guidelines for Preparing Proposals* is the best-selling title ever published by Lewis Publishers/CRC Press. At this writing, it had sold over 15,000 copies. It also has been selected by a business book club and used in various college classes. Meador has recently published a second edition. In addition, Meador has published hundreds of articles on a wide range of subjects in such national periodicals as *The New York Times, Smithsonian, Chess Life, True West,* and *Analog.*

Of the six survey respondents, Meador provided the most comprehensive response. His remarks cover $8^1/_2$ single-spaced pages. Here are some excerpts.

About his writing career: "My first published book was *Future Energies*, published by Ann Arbor Science Publishers in 1974. The book, a response to the energy crisis of the 1970s, examined the energy alternatives to fossil fuels including solar energy, fusion power, geothermal, tidal, and water powers. The book valiantly proclaimed that it would take 'a right now look at our energy prospects for tomorrow...and a thousand years from tomorrow.' Operation Desert Shield and the oil guzzling 1980s and 1990s demonstrate that the arguments of the work have yet to take hold in a still energy-illiterate America.

"My most significant technical publications are *Franklin-Revolutionary Scientist* and *Guidelines for Preparing Proposals*. The first was important to me because it caused me to delve deeply into the remarkable life of the inimitable Ben Franklin, launching an inexhaustible and lifelong enterprise of informative, rewarding, and pleasurable study. [The second book has meant] most in terms of impact, sales, and readership."

His advice to would-be technical writers: "[W]hat does it take to get into print? Persistence, thick skin when the cascade of rejection slips arrive, and more persistence... The first person a writer has to sell is not an editor or a publisher, but himself. He has to believe that he has something worth saying, and he has to keep saying it with those little words on paper until the right combination occurs and somebody somewhere listens and publishes... Yes, most good writing is very hard [work], but being published makes up for the hard work. And when you get paid to boot... ah, paradise enow.

"I began writing for pay at 15 on a small town newspaper. I worked as a printer and proofreader in the days before computers and their treacherous 'Spell Checkers' rendered whole generations insensitive to correct English. I learned the rudiments of writing decent, reasonably correct, accurately spelled prose. That era was what we now call 'the good old days' before language pollution and deterioration became fashionable. Now the phrase could appear in print as 'the gud olld daze' and most readers wouldn't see anything wrong."

How does one learn what it takes to get into print? "I can't answer the question for *you*. I doubt anyone else can. Newsman O.O. McIntyre said it pretty well long ago: 'I learned to write by writing. I don't believe there is any other way.' I don't believe there is either. Courses in schools help a lot mainly because there are writing assignments that have to be suitably accomplished for a passing mark. But there are no 'secrets' that I've discovered... If you want to write, write. Then fix it up and send it out. And keep writing on something new while you wait for the piece to come back, or glory be, a check! Scribble, scribble, scribble! That's the formula for writing success and the only one."

Are there any references no writer's library should be without? Meador considers the following books indispensible: *Webster's Ninth New Collegiate Dictionary, The Encyclopedias Brittanica* and *Americana, The Columbia Encyclopedia, Roget's Thesaurus of English Words and Phrases*

(C.O. Sylvester Mawson), *Roget's Thesaurus of the English Language in Dictionary Form* (Mawson), *World Almanac* (current), several books of quotations, and several stylebooks and guides. In addition, he has amassed a library of 10,000+ books "in all fields — scientific, technical, literature, history, general, and growing. All of these...are source and resource volumes for my work."

On his writing habits: "For me, the quietest time of day is the best time for writing. So I tend to start pretty early, before sunrise. When I'm on a project for a client, 12-hour days are the norm. Sometimes 14- and 16-hour work days take over in the second or third week of a big job as hearty involvement and/or obsession to finish take charge of the writing process. I am not much of a believer in pleasure or relief 'breaks' while a writing task is underway. I prefer the peaceful break that follows completion of an assignment." Nor does he believe in writing in spartan surroundings: "Yet certain therapeutic indulgences are permissible and important as part of the writing environment. Background music, for instance, helps me combat noise distractions. Also, coffee, hot and constant, of course, is an essential catalyst of composition."

Having used typewriters for many years, Meador now faces "the screen of a personal computer and hope[s] for the best. No, on that computer I do not use any 'Spell Check' except myself and the dictionary. Correctness is the writer's responsibility, not [that of] some inconsistent electronic whizbang."

Before he starts writing, he spends time "putting order into the [material] that will supply data for the work ahead. The more care I take during this organizing, the easier and faster I find the subsequent writing job. When working with some formidable and chaotic mountain of technical information, the ability to lay hands promptly on any item needed is the hoped-for dividend of advance organizing efforts. Such attempts to put your paper house in order before starting may seem boring and time consuming, and you may itch with impatience to start writing. Believe me, it is not a waste of time."

And Meador believes in staying at a writing job until it's done: "While writing, I usually have a working lunch. I've heard of 'coffee breaks', but I don't think that's a respectable, profitable, or even possible freelance practice. Sometimes when stuck about how to proceed, weather permitting, I take a walk. Casual walking alone is an effective thought-arranging and idea-organizing exercise for writers." (Or for anyone else, for that matter!) However, "[s]elf-discipline is the toughest trick of all. I find it easiest to make myself buckle down and crank it out when a client gives me a deadline I need and want to meet, because that's when I get paid. A lot of my writing is done on an hourly fee basis. Professional integrity makes me try to see that each hour is honestly productive, which is another effective means of achieving self-discipline."

Finally, "Every writer learns the necessity of a 'Don't Dare Disturb

Me!' sign surrounding him like a cocoon. Silence, solitude, and no interruptions are necessary compulsions."

What about the technical professionals-can't-write-syndrome? "I love this misconception, of course. It has brought me a lot of writing assignments. Writing as art is a talent that some have and most don't; but writing as a communications skill can be learned by most technical professionals — if they study, practice, and work at it... *Tip*: If you write today and write tomorrow, day after tomorrow you'll find it goes easier than today. But if you wait a month to try again, you'll be back at square one and have to sweat out the perpetual struggle involved in getting started.

"I suspect that most professionals who avoid writing are simply avoiding the hard and lonely work involved. And let's not kid ourselves, hard it generally is, lonely it always is... I imagine most technical professionals realize they can write; but they simply don't want to lock the door, banish visitors, turn on the answer machine, and spend the next three days sweating out the words...Meetings, lunches, networking...they're apple pie by comparison."

His philosophy of writing: It is "the same as my philosophy of climbing mountains: Good writing and good climbing emerge from hard working; you don't reach the 'top' or the 'end' by just talking or thinking about it, you have to tackle the task and fight your way through to the finish. The secret of effective writing that communicates and sells is locked up in these not-so-secret words: Reading, Research, Notes, Organization, Order, Editing, Rewriting, Proofreading, Tenacity, Persistence."

Meador's final words on writing and publishing: "I think we have to consider each publisher as an individual and pay close attention to what he wants if we want his money and expertise for our work...If you find a publisher who seems promising for your sort of thing, cultivate him, learn his likes, dislikes, needs; and act accordingly."

Henry Petroski

Few writers, technical or otherwise, have accomplished as much in their trade as Henry Petroski has. His writing has earned him respect and acclaim among both academics and the general public. Moreover, his work has bridged the gap between the technical and everyday worlds. A New York City native, Petroski earned his undergraduate engineering degree from Manhattan College and his masters and Ph.D. degrees from the University of Illinois at Urbana-Champaign while there on a teaching fellowship. His first research paper won an award from Sigma Xi, the scientific research society. After teaching engineering mechanics at the University of Texas at Austin and doing research work at the Argonne National Laboratory, he joined the Duke University engineering faculty in 1980. As a Professor of Civil Engineering, he carries out research in such areas as fracture mechanics and structural dynamics. Also at Duke, Petroski has served as director of civil and environmental engineering graduate studies and has participated in various interdisciplinary activities, including developing a structural engineering course for non-engineering students. The latter activity has evolved naturally from his long-standing interest in broad issues of technology and society and in the relationship of engineering to our general culture.

This interest has impelled him to write — not just 60+ refereed journal articles in his specialty, but also poetry and essays for publications as diverse as *Technology Review, Descart, Issues in Science and Technology*, the *New York Times*, and the *Washington Post*. He collected some of these essays in his book, *Beyond Engineering*.

In 1985, St. Martin's Press published Petroski's *To Engineer is Human: The Role of Failure in Successful Design*, a book that has been well received by engineers and non-engineers alike, both here and abroad. In 1987, the British Broadcasting Corporation broadcast as part of its *Horizon* series a television program based on this book. Narrated by Petroski, the program has been broadcast worldwide, including over PBS stations in the U.S. and Canada.

During 1987–88, while on fellowships from the National Endowment for the Humanities and the National Resources Center, Petroski wrote *The Pencil: A History of Design and Circumstance*. Published by Alfred A. Knopf in January 1990, *The Pencil* has attracted considerable attention for its extended treatment of the history of the common pencil as a means of explaining engineering and its interaction with society and culture. A recipient of a 1990–91 Guggenheim Fellowship, he used the year to work on a new book about how engineering shapes the everyday world.

Petroski's views on technical writing (written in pencil, naturally) are terse but thought provoking:

How to write well and get published? "I believe I have learned most about writing by reading other writers. And practice may not make perfect, but it makes better, I think."

About his writing habits: "[I adhere to] as regular a schedule as possible, subject to my other obligations."

Rather than express an opinion about the technical professionals-can't-write misconception, he observes: "This can become a self-fulfilling prophecy."

Finally, his "philosophy of writing" is characteristically candid: "I'd prefer not to attempt to summarize something I'm not sure about."

THE EDITORS AND PUBLISHERS

Harold M. Englund

"Hal" Englund has been in the technical writing field since, as the saying goes, "Hector was a pup." His career spans over 4 decades — a period which has seen much growth and many changes in the business. However, this is not to say that either Englund or his ideas are by any means old. Quite the contrary. He is an energetic, positive-thinking gentleman with a receptiveness for new ideas that many men half his age fail to possess.

Hal's career began in 1950 with his graduation from Newark College of Engineering with a B.S. in mechanical engineering. He first spent 7 years as a plant engineer with T.A. Edison in West Orange, New Jersey. Following that, he took a position with Industrial Press (New York City), as an assistant editor for the magazine, *Air Conditioning Health and Ventilating*. After 7 years with this publication, during which time he advanced to become its managing editor, he left in 1965 to become editor of *JAPCA*, the Pittsburgh-headquartered *Journal of the Air Pollution Control Association* (now *JAWMA* — *Journal of the Air and Waste Management Association*). He has been there ever since. During this quarter century, he has seen the journal "grow from a 64-page monthly dealing with air pollution control to a 120-page publication covering additional environmental areas, such as waste management..."

Though by trade a technical editor, Hal's engineering credentials are nonetheless impressive. A former student at the University of Pittsburgh Graduate School of Public Health, he is a registered professional engineer in Pennsylvania and New Jersey. He is also a Diplomate in the American Academy of Environmental Engineers. Nor has he limited his accomplishments to editorship of *JAWMA*. He co-edited and contributed to the *Handbook of Air Pollution Technology* (John Wiley & Sons, 1984). This handbook has been well-received domestically and abroad and, in 1980, was translated into Russian by the *Metallurgical Institute of Moscow*.

During his many years as an editor, Hal has read a great deal of writing, both good and bad. Unsurprisingly, he has formulated some ideas on the art of written expression. His advice to anyone wanting to get published: "Do an outline first! Have [this] outline approved by [the] editor/publisher." And once this outline is approved, the writer should "use the ACS (American Chemical Society) style guide" when writing. Another bit of advice: "Write to inform — not to impress!"

Do many technical professionals have trouble writing? Hal thinks so: "Many *cannot* write, and they should get help before submitting a manuscript to a publisher. I'm often shocked at the quality of writing (sentence structure, grammar, etc.) that I receive as a 'finished' manuscript."

His final comment on the technical writing business is characteristically upbeat: "Being an Editor (capital 'E') is the best and most rewarding job there is!"

Richard A. Young

Richard A. ("Dick") Young is the founding editor and publisher of *Pollution Engineering* (PE) magazine. He has brought a lifetime of professional experience to PE, having served as an environmental advisor and consultant to private industry, 14 states, five federal agencies, and more than a dozen municipal governments. Young was also instrumental in the founding of the Water and Wastewater Industry Division of the Instrument Society of America and helped establish the first national meeting of environmental equipment manufacturers. He has also served as an adjunct professor of environmental engineering at several major universities. Finally, as an experienced design engineer of air and water pollution control equipment, he holds 11 patents and claims.

Having spent several decades in technical publishing, he has experienced both the good times and the bad. His assessment of the current situation leans more toward the latter pole: "The recession and changes in postal regulations have vividly brought to light some of the problems of the technical publishing business today. Magazines are up for sale as well as entire publishing companies. What will ultimately result will be fewer, but stronger publishing companies."

Though he's an editor and publisher, I often think that Dick Young is, like many of us, a writer at heart. His monthly "Viewpoint" column in PE is light and humorous, a pleasure to read. In fact, when I pick up the magazine, his column is the first thing I read. I think his views of technical editing — and writing — reflect this:

"As a technical editor, you learn to improve your editing abilities every day. With each piece you produce, there will always be someone who writes to you with their critical comments. Writing is something I have had to learn — and it hasn't been easy. I was hired as an editor because I was an engineer. My boss told me that he could take an engineer off the streets and teach that person to be a good editor of other people's work in 6 months, but he couldn't take a journalist off the street and teach him engineering in less than 4 years. While I would disagree with him now about time frames, it was an interesting statement that provoked me into learning new skills and thought processes. I attempt to write something every day.

"The editor must be able to get into the author's head and determine why something has been stated in the manner in which it has. The editor must be able to understand everything about why and for whom the material has been written by the author, if it makes sense and gets to the point."

However, like most technical editors, he has found that technical professionals have difficulty writing articles. In Young's opinion, this is because they "learn to write reports, not manuscripts for publication. It is a difficult learning experience for the recent technical graduate to have a report rejected by an employer because it is not present in a report-type manner. Breaking in the technical professional is not an easy task once this lesson has been learned. Take it from one who knows and has worked on both sides of the drafting table."

His closing remark sums things up rather nicely: "My philosophy of editing was taught to me by my father, 'Never be satisfied'!"

Jeffrey G. Hillier

A native of Great Britain, Jeffrey Hillier was educated there, receiving his B.Sc. (Applied Biology) and Ph.D. (neuropharmacology) degrees from the University of Bath. After a stint as a research scientist with Aspro Nicholas (a British pharmaceuticals firm), he took a position as an acquisitions editor with Elsevier Science Publishers, Amsterdam in 1975. He advanced through the Elsevier ranks, becoming first a senior editor, then Deputy Managing Editor, Managing Director of the Biomedical Division, and finally Director of New Business Development in 1986. After a year as Editorial Vice President of Frost and Sullivan, he moved to his present position as Vice President/Editor-in-Chief of CRC Press (Boca Raton, Florida) in 1990.

Hillier's questionnaire responses are thoughtful and carefully considered:

What is his overall assessment of the technical publishing business today? "[It is] very difficult to generalize. Scientific and technical publishing is international, and market conditions vary according to national and regional economics. If forced to answer, I would say healthy and growing slowly."

"Healthy and growing slowly." Those four words describe the state of technical publishing in a nutshell. Technical publishing has not (nor, some think, *should* it) exhibit alternating periods of boom and bust, as do some corners of the trade publishing world. Conversely, technical publishers, much like the authors who write for them, are a conservative, cautious lot who try to achieve steady, respectable sales on a long list of titles that they keep in print two, three, or more years. Unlike trade houses, who invest (gamble) large sums of advance monies on the hope that a tiny number of books will become bestsellers, technical publishers spread their risks over several books and several years.

Yes, technical professionals are conservative, by and large. But are they poor writers, as well? "True," replies Hillier, "[but] many people do not write well. However, this should not deter them from making the effort. We employ editors to brush up their style and grammar. In technical publishing, the ability to present facts in context is very important."

This last sentence reminds me of what another technical book publisher told me. This person (who asked not to be identified) said that for technical professionals, engineers and scientists primarily, the ability to write well is secondary to the ability to collect and present useful and important facts clearly and comprehensively. "We have copy editors to clean up their prose. However, *no* copy editor can formulate technical theories and collect the data to substantiate them. That is what our authors do. If they can write, fine; if not, fine too. The writing ability is not that important."

This publisher also said that there is a distinct difference between a technical professional who writes and a technical writer. The former would be an engineer, scientist, etc., who records her thoughts and findings, but who does not write for a living. On the other hand, a technical writer is one who, like Bob Bly and Roy Meador, earns his bread from what he writes. He can have a technical background, but it isn't a requirement. A technical writer can write any number of things: books, advertising copy, technical reports and manuals, ghost-written pieces for corporate executives, or anything else connected with the technical world. Finally, this publisher identified a third category — a journalist who specializes in technical topics, such as the science writers whose work appears in popular magazines and newspapers.

Though interesting to consider, these distinctions are somewhat academic, at least as far as this book and *you* are concerned. For I happen to believe that if you — technical professional, technical writer, or journalist — write technical prose, you share a goal with all other writers. And that goal is to get your work into print. This book, I hope, has made and will make that goal easier and more enjoyable to attain.

INDEX

DATE DUE

SEP 24 '94			
DEC 19 '96			